AF763550

NATO Science for Peace and Security Series C: Environmental Security

NATO Science for Peace and Security Series
Sub-Series C. Environmental Security

Volumes in the Environmental Security Sub-Series deal with scientific and technological aspects of innovative and promising projects related to environmental security, including regional environmental problems and natural and man-made disasters. Floods, droughts, landslides, desertification, earthquakes and industrial accidents are examples of topics covered, as are regional environmental problems related to transboundary water resources or transfrontier air pollution. Other areas covered relate to climate change, environmental chemistry, disaster prevention and forecast, remote sensing and detection, and decision support systems/risk assessment.

Special offer: For all clients with a print standing order we offer free access to the electronic volumes of the Series published in the current year.

Maria Fabrizia Buongiorno · Enrico Casini ·
Andrea Manciulli
Editors

The Challenge of Environmental Security in the Euro-Mediterranean Region

Strategic Implications, Crisis Management, and Technological Applications

Springer

Editors
Maria Fabrizia Buongiorno
Istituto Nazionale di Geofisica e Vulcanologia (INGV)
Rome, Italy

Enrico Casini
Med-Or Italian Foundation
Rome, Italy

Andrea Manciulli
Med-Or Italian Foundation
Rome, Italy

ISSN 1874-6519 ISSN 1874-6543 (electronic)
NATO Science for Peace and Security Series C: Environmental Security
ISBN 978-3-031-99645-0 ISBN 978-3-031-99646-7 (eBook)
https://doi.org/10.1007/978-3-031-99646-7

Proceedings of the NATO ARW on Climate Change and Natural Hazards in the Euro-Mediterranean Region: Security Impacts and Crisis Management Rome, Italy 19–21 November 2024

© The Editor(s) (if applicable) and The Author(s), under exclusive license to Springer Nature Switzerland AG 2025

This work is subject to copyright. All rights are solely and exclusively licensed by the Publisher, whether the whole or part of the material is concerned, specifically the rights of translation, reprinting, reuse of illustrations, recitation, broadcasting, reproduction on microfilms or in any other physical way, and transmission or information storage and retrieval, electronic adaptation, computer software, or by similar or dissimilar methodology now known or hereafter developed.
The use of general descriptive names, registered names, trademarks, service marks, etc. in this publication does not imply, even in the absence of a specific statement, that such names are exempt from the relevant protective laws and regulations and therefore free for general use.
The publisher, the authors and the editors are safe to assume that the advice and information in this book are believed to be true and accurate at the date of publication. Neither the publisher nor the authors or the editors give a warranty, expressed or implied, with respect to the material contained herein or for any errors or omissions that may have been made. The publisher remains neutral with regard to jurisdictional claims in published maps and institutional affiliations.

This Springer imprint is published by the registered company Springer Nature Switzerland AG
The registered company address is: Gewerbestrasse 11, 6330 Cham, Switzerland

If disposing of this product, please recycle the paper.

Foreword

Strengthening Cooperative Security Through NATO's Science for Peace and Security

At the NATO Summit in Madrid in 2022, Allied Heads of State and Government adopted a new Strategic Concept that marked a decisive step forward in the Alliance's approach to partnerships. This renewed vision placed greater emphasis on supporting partners to enhance their ability to address shared security challenges, underscoring the value of collaboration in an increasingly complex global environment.

Since then, NATO has progressively scaled up its capacity-building initiatives. The most recent milestone came in the summer of 2025, when Allied leaders, gathered at the NATO Summit in Washington, endorsed the Southern Neighbourhood Action Plan. This plan reflects a stronger, more strategic, and results-oriented approach to engagement with the region. It recognises that the interconnected challenges present in NATO's southern flank, exacerbated by an increasingly complex regional situation, and directly affect the security of Allies partners alike. The plan also acknowledges that NATO's southern neighbourhood is a source of opportunities for dialogue and cooperation in areas of mutual interest. In this context, partners in the Middle East and North Africa (MENA) play an essential role as active contributors to regional and international security. In this context, cooperation within the Science for Peace and Security (SPS) Programme proves particularly valuable.

The SPS Programme is a central component of NATO's science and technology ecosystem. With a legacy spanning over 60 years, SPS has helped to demonstrate NATO's commitment to scientific and technological advancement, by supporting hundreds of collaborative research and knowledge-sharing initiatives on security-related issues. The topics addressed by SPS deal with new, contemporary challenges, such as innovation and emerging and disruptive technologies, cyberdefence, resilience, energy security.

What these topics have in common is that they are often multidisciplinary and transboundary in nature. And to deal with these challenges, cooperation is the key. This is why NATO and its partners shall continue to work together, in the spirit of cooperative security, through the SPS Programme. The SPS Programme has a unique way to make this happen: engaging scientists. In its activities, they explore solutions to challenges of mutual concern, on a local, regional and global scale. In doing so, SPS has established itself as one of our most valuable partnership programmes. And throughout history, it has proved to be a highly effective medium for international dialogue and for fostering enduring collaborations.

This publication is the outcome of a workshop convened under the SPS Programme, dedicated to addressing pressing regional challenges. The findings and recommendations that emerge from this work will inform and strengthen NATO's future engagement in these critical areas.

Claudio Palestini
Head, Innovation and Technology Adoption
NATO
Brussels, Belgium

Acknowledgements

This publication has been made possible through the generous support of the NATO Science for Peace and Security Programme. Several of the texts included herein constitute the outcomes of research and studies conducted by speakers in the workshop held in Rome from 19 to 21 November 2024.

We wish, indeed, to express our profound gratitude to the NATO SPS Programme, the Italian National Institute of Geophysics and Volcanology (INGV), and the Ministry of Foreign Affairs of the Hashemite Kingdom of Jordan for their indispensable contributions.

Our sincere thanks also extend to the Med-Or Italian Foundation and as well as to all the experts, scientists, and researchers whose insights and work have been vital to the realization of this scientific publication.

We are profoundly grateful to Massimo Khairallah, Lucrezia Di Noia, and Emanuela Manganelli for their invaluable assistance in organizing the November event, as well as to Federico Deiana and Manfredi Martalò, whose efforts in editing and refinement have significantly enhanced the editorial process and the compilation of materials presented in this volume.

Conflict of Interest This publication has been made possible through the generous support of the NATO Science for Peace and Security Programme.

Several of the texts included herein constitute the outcomes of research and studies conducted by speakers in the workshop held in Rome from 19 to 21 November 2024.

The opinions expressed by the authors in this publication are solely their own and do not necessarily reflect the views of the editors.

Introduction

Historically, the primary threat to the security of a territory, a state, or its inhabitants has been represented by various forms of violence and armed conflicts. Throughout human history, the security of individuals, populations, and their social organizations has often been jeopardized by factors and events of natural and environmental origin. Reviewing the various historical epochs, from prehistory onward, it becomes evident that, on numerous occasions, natural, climatic, and geological events have severely impacted or even threatened the survival of entire inhabited regions. Some of these events have even become foundational elements of ancient legends and myths, shared by multiple cultures. For example, the myth of the Great Flood or the legend of the destruction of Atlantis.

Moreover, human history has largely, at least until the Scientific and Industrial Revolutions, been a continuous struggle for survival, strongly influenced by nature and its manifestations, even the most extreme ones. It is no coincidence that in ancient times, many deities were associated with atmospheric or natural phenomena, or the disasters that could befall a community or territory were attributed to the wrath of the gods.

From the Industrial Revolution onward, and especially due to the significant technological advancements of the 20th century and the early decades of the 21st century, the very concept of security has assumed an increasingly expansive and global dimension. Technological revolutions, coupled with economic globalization, have been pivotal accelerators in the expansion of the security field. In an increasingly interdependent and connected world, where events occurring in one place can have far-reaching effects, and at a time when humanity operates across multiple domains—surpassing the physical barriers of land, air, and water, and even venturing into space or the virtual realm—it is evident that the context of global interaction has profoundly changed. The technological revolution, which has equipped us with increasingly advanced tools and capabilities—from nuclear energy to genetics to artificial intelligence—has further expanded the scope of security, giving it a breadth and complexity that now influences every domain of human activity.

As history has taught us, when the security of a state, territory, or community is threatened, the stability and continuity of its institutions, which are essential to ensuring well-being, social peace, and political unity, can also be undermined. As previously mentioned, a significant portion of the threats to the security and stability of peoples and their social organizations throughout history has come not only from wars, revolts, or invasions, but from natural and climatic events: floods, earthquakes, volcanic eruptions, radical climate changes, and droughts. Climate has always been one of the major factors conditioning life on our planet, and, of course, the conditions under which humankind has spread and colonized lands and continents. But, in turn, the climate has been influenced by geological events that have occurred over millions of years or by great natural disasters that have happened suddenly. From massive volcanic eruptions to tectonic movements, to the impact of asteroids or meteorites, which may have conditioned the development of life, as in one of the theories about the extinction of the dinosaurs. The environment, in general, has been decisive for the development of human civilizations; the often-conflictual relationship with nature has represented a constant throughout humanity's evolutionary path.

The great climatic stability that followed the end of the last Ice Age allowed the development of agriculture, and with the Neolithic Revolution, humanity transformed from hunter-gatherers to farmers, establishing a presence in certain regions of the planet. Agricultural development led to the birth of the first civilizations along major rivers, primarily based on cereal cultivation. Specifically, in the alluvial plains of the Nile, the Tigris and Euphrates, the Indus, and the Yellow River, the first structured forms of civilization emerged, followed by other similar sedentary cultures that built the first cities. Yet, just as it had been for cavemen or for hunter-gatherer communities during glaciations, the relationship with nature and the risks arising from the environmental conditions in which these communities settled always represented both a factor of development and a potential risk. The combination of periods of climatic instability and natural events, such as a Nile flood smaller than needed, could lead to a bad harvest, resulting in famine and, consequently, political turmoil, crises, or revolts. Similarly, an earthquake, an unexpected flood in monsoon-prone areas, or catastrophic volcanic eruptions could cause great damage to human populations. Human existence is, therefore, strongly influenced by risk factors associated with natural events. Kingdoms and empires have collapsed throughout history due to abrupt climatic changes that triggered crises, mass migrations, and conflicts, often in combination with other unforeseen natural manifestations. Consider, for instance, the case of the Akkadian Empire, which likely collapsed due to a long period of climatic crisis and severe drought, or the cataclysms that struck Thera and Crete during the Minoan era.

Some historical periods, marked by significant temperature variations—often characterized by a cooling of the climate in certain regions—were frequently triggered by violent volcanic eruptions, even from distant locations, such as those in 19th-century Indonesia (the Krakatoa eruption in 1883 and the Tambora eruption in 1815). The eruption of Tambora produced a terrible famine in Europe, which also caused significant emigration to the USA, where the eastern coast experienced the same effects of the eruption. The resulting impact on the environmental, socio-economic, and political conditions of many human societies was often severe and lasted for long periods. Some more recent cooling phases, for example in late antiquity or the modern era, were probably due to major volcanic eruptions.

These natural events, alongside many other factors of various origins, constitute the broad spectrum of risks and threats that belong to the realm of environmental security, one of the most significant areas in the vast global security landscape today. In our time, amidst the complexity of the world in which we live, the fragility of supply chains, the importance of critical infrastructures, agricultural or industrial production chains, and the levels of our complex social, economic, and political system—all exposed to environmental risks—are numerous and often interconnected, transnational, and cross-cutting.

Environmental security, indeed, is an issue that requires a cross-cutting approach, which knows no borders and, by its very nature, deals with threats and events that can affect regions and areas of the planet, transcending the borders of states themselves. The more our society becomes global and interconnected, the greater the risks arising from natural and environmental events have expanded their potential for causing damage in diverse areas. Today, the scope of security has extended to a plurality of sectors. The impact of catastrophic events on supply chains, critical infrastructures, the economy, technologies, and cultural heritage has grown proportionally to the levels of interconnection and interdependence. Consequently, the risks linked to climate change have also increased. In recent years, the concept of climate change as a "crisis multiplier" has become more widely accepted.

Consider, for instance, areas affected by prolonged periods of drought. In such regions, water scarcity can undermine agricultural and livestock sectors, causing severe economic damage and food crises. However, it can also become a trigger, particularly in more unstable or poorer countries, for social tensions, revolts, or disorder, and may even foster the rise of phenomena such as migration or terrorism. Furthermore, the competition for control over water resources or fertile territories could lead to forms of rivalry that may escalate into genuine conflicts—both interstate and intrastate (e.g., among different social, political, or ethnic groups).

The same geopolitical competition could see an increased importance of environmental and climatic factors in global or regional dynamics. In the Arctic, for example, the melting of glaciers will make it easier to exploit the region commercially and for extractive purposes, given the new shipping routes passing through the Arctic Sea and the vast quantities of oil, gas, and minerals contained within it. Drought, desertification, and demographic growth could increase competition among countries for important water reserves, as demonstrated by the dispute between Ethiopia, Egypt, and Sudan over the construction of the GERD dam on the Nile River. Similarly, in certain critical contexts, these issues might encourage recruitment to insurgent or terrorist organizations, as seen in some areas of Sub-Saharan Africa.

Lastly, current climatic and environmental events may have serious consequences in the medium-to-long term. For example, the risks associated with rising sea levels, stemming from the ongoing melting of polar ice. Over various historical periods, sea levels have fluctuated around the planet. What could happen today to ports, strategic infrastructures in coastal areas, naval military bases, and energy-extracting platforms if sea levels rise in the coming years? What consequences might there be for human communities living in at-risk coastal zones? These are only a few of the questions that must be raised concerning the risks associated with climate change—without descending into excessive catastrophism, but by carefully considering the scientific data currently at our disposal.

Thus, growing importance of environmental security for global security is becoming increasingly clear. For these reasons, it is essential to address the topic not only in terms of prevention, damage reduction, or mitigating the effects of natural risks, but also in managing the emergencies and crises that may arise. These are complex challenges that cannot be tackled by a single individual or sector, but will require, both nationally and internationally, cooperation and collaboration at the political and diplomatic levels, as well as in the economic, scientific, and technological spheres. Approaches to these challenges should include studying field experiences and employing advanced technologies, from digital to space technologies, to provide useful lessons and tools to address them.

In the near future, it is likely that new technologies and scientific research will enable us to meet the challenge posed by natural disasters. Environmental security is central for states and international organizations, and it is therefore essential to strengthen reflection and study on these issues. Based on the experience of our workshop organized in November 2024 by the Med-Or Italian Foundation, the Ministry of Foreign Affairs of the Kingdom of Jordan and the National Institute of Geophysics and Volcanology (INGV), with the support of the NATO SPS Programme, we have realized this publication, with contributions from several experts in these fields, some of whom participated in the workshop's discussions.

This collection of essays aims to offer a useful contribution to raising awareness about these issues, also through the study of some field experiences and specific case studies. It will also provide a foundation for reflection on future cooperation and scientific research initiatives that the Med-Or Italian Foundation, in collaboration with other national and international partners, can promote.

Our publication includes contributions from numerous experts who have addressed the impact of natural threats and climate change on security from various perspectives, offering analyses of specific case studies and possible intervention and prevention tools, particularly through the use of modern digital and satellite technologies available in this field.

The contributions cover topics directly related to climate-related events and natural disasters. In some cases, the economic, social, and geopolitical impacts of these events have been evaluated, as well as the direct threats posed to critical and strategic infrastructures. Particularly in the second part of the publication, several case studies are presented.

Given the vastness of the topic, it was impossible to exhaustively cover all the different types of threats and risks or to include a complete set of examples and studies that could answer every possible scenario. We have focused on particularly relevant themes, especially for the Euro-Mediterranean area, which could offer starting points for reflection and further exploration of the issue.

The region's geographical, geological, and environmental characteristics, combined with the socio-political conditions of many of its countries, make it particularly vulnerable to these types of natural, environmental, and climatic risks. It is evident that national and international institutions involved must foster cooperation and collaboration to prevent these risks and manage crises stemming from them. By doing so, they can limit the impact and damage and reduce the consequences that might lead to increased insecurity and instability in a region already severely affected by conflicts and tensions but also increasingly strategic for the global economy, trade, and politics.

Our hope is that this work, developed with the contributions of numerous experts and scholars, whom we sincerely thank, and with the support of the NATO SPS Programme, can offer a valuable contribution to the study and reflection on these increasingly important issues.

Maria Fabrizia Buongiorno
Enrico Casini
Andrea Manciulli

Contents

Editors and Contributors

About the Editors

Maria Fabrizia Buongiorno is the Director of Technology Research at the Italian National Institute of Geophysics and Volcanology (INGV) and Coordinator of INGV Centre for Earth Observation from Space (COS). Graduated in Geology at University of Rome "La Sapienza" in 1988, she has more than 30 years of experience in the sector of EO missions and applications with particular focus on environment and natural hazards. She started her career in Telespazio company in the Earth Observation Division. In 1992 she moved to the Istituto Nazionale di Geofisica (ING), where she support the development of a research area dedicated to remote sensing techniques. Since 2003, she is Director of Technological Research at Istituto Nazionale di Geofisica e Vulcanologia (INGV) and responsible of INGV Remote Sensing Activities from 2004. From 2020 (current) she is Coordinator of the new INGV Centre dedicated to Space Observations of Earth (COS) which includes EO and Space Weather application and services. From 2021 to 2024, she has been appointed by the Italian Space Agency, President of E-GEOS Company (a Telespazio/ASI company). 2018–2024 She was Administrative Responsible for a large infrastructural project funded by Ministry of Research (MUR) which renewed and enlarged the INGV field networks and in laboratories and activated more than 40 post doc fellowships. In the period 2017–2024 she has been a member of the National Operative Committee for the Civil Protection to represent INGV. She worked in close cooperation with aerospace industries to propose new EO mission concepts which may improve the understanding of Earth Processes and Climate Change. She is coinvestigator with Dr. Simon Hook of Jet Propulsion Laboratory and member of the ASI Advisory Board in the development of NASA-ASI SBG-TIR mission (Surface Biology and Geology, TIR element), and member of other Advisory Boards for European space missions with Optical payloads for Earth Observations (ESA-LSTM, ASI-PRISMA, DLR-EnMAP). Since 2013, she is co-chair with USGS of the Earth Science Working

Group in the frame of the USA-ITALY bilateral agreement in Science and Technology and coordinates the activities in the MAECI-INGV MoU. She is author of a large number of papers and publications in the above-mentioned fields.

Enrico Casini is the Director of Communications of Med-Or Italian Foundation. Graduated in International Relations at the University of Siena, he is an expert in communication, strategic and political analysis, institutional relations, and event organization. During his career he has gained a wide experience in the institutional field, both at the national and international level. He has been Advisor to the Minister of Defense, and he has collaborated, among others, with the Regional Council of Tuscany, the Chamber of Deputies of the Italian Parliament and the NATO Parliamentary Assembly. From 2020 to 2021, he oversaw NATO-EU geo-strategic relations at the Strategy and Market Intelligence Office of Leonardo S.p.A. Since July 2021, he has become Head of Med-Or Italian Foundation Communications. In his professional career, he also produced publications and articles on magazines, blogs and journals on international terrorism, security and US-Europe relations. He is co-editor of *Il futuro del terrorismo di matrice jihadista* (published by Ledizioni), *Vent'anni di Guerra al terrore* (published by START Insight), and *La guerra tiepida. Il conflitto ucraino e il futuro dei rapporti tra Russia e Occidente* (published by Luiss University Press).

Andrea Manciulli is the Director of Institutional Relations of Med-Or Italian Foundation. A Modern History graduate from the University of Pisa, he studied at the École des hautes études en sciences sociales (EHESS) in Paris and is an expert in military history and international terrorism. He was a member of Tuscany's regional council from 2000 to 2013, then went on to become a Member of the Italian Parliament (2013–2018) and served as Vice-President of the Parliamentary Commission "Foreign and Community Affairs" and President of the Italian Delegation to the NATO Parliamentary Assembly. While serving in this position, he was appointed as President of the Special Mediterranean and Middle East Group (GSM) of the NATO Parliamentary Assembly. While at the NATO Assembly, he curated reports on jihadist terrorism (2015–2017). In 2015 he was the Italian Chamber of Deputies' rapporteur for the Decree on "Urgent measures to counter terrorism" adopted by the Government on 18 February 2015 and converted into law on 17 April 2015. In January 2016, in this role, he also presented, together with Stefano Dambruoso, a bill on "Measures for the prevention of radicalization and jihadist extremism". In 2016–2017 he was rapporteur of the Framework Law on International Missions. From 2018 to 2020, he was Executive at Fincantieri S.p.A. and President of the Fincantieri Foundation. From 2020 to 2021, he was Senior Vice President in the Strategy and Market Intelligence Office of Leonardo S.p.A. He is the President of the think tank Europa Atlantica and the author of several publications on jihadist terrorism, national security, and international politics.

Contributors

Ruba Ajjour Climate Change Studies Division, Royal Scientific Society, Amman, Jordan

Jawad Al-Bakri School of Agriculture, University of Jordan, Amman, Jordan

Muhammad Rasool Al-Kilani School of Agriculture, University of Jordan, Amman, Jordan;
School of Natural Resources Engineering and Management, German Jordanian University, Amman, Jordan

Tommaso Alberti National Institute of Geophysics and Volcanology (INGV), Rome, Italy

Marco Anzidei National Institute of Geophysics and Volcanology (INGV), Rome, Italy

Simone Atzori National Institute of Geophysics and Volcanology (INGV), Rome, Italy

Maria Fabrizia Buongiorno National Institute of Geophysics and Volcanology (INGV), Rome, Italy

Sandro Carniel Institute of Polar Sciences, National Research Council, Venice, Italy

Enrico Casini Med-Or Italian Foundation, Rome, Italy

Massimo Chiappini National Institute of Geophysics and Volcanology (INGV), Rome, Italy

Stefano Corradini National Institute of Geophysics and Volcanology (INGV), Rome, Italy

Giorgiana De Franceschi National Institute of Geophysics and Volcanology (INGV), Rome, Italy

Federico Deiana Med-Or Italian Foundation, Rome, Italy

Carlo Doglioni Department of Earth Sciences, La Sapienza University, Rome, Italy

Marco Dordoni University for Foreigners, Perugia, Italy

Andrea Gilli School of International Relations, University of St. Andrews, St. Andrews, Scotland, UK

Stephane Guillot University of Grenoble Alpes, University of Savoie Mont Blanc, CNRS, IRD, UGE, ISTerre, Grenoble, France

Tanguy Jonville Sorbonne University, CNRS, LATMOS, Paris & Ecole Des Ponts, Marne La Vallée, France

Carlo Marcocci National Institute of Geophysics and Volcanology (INGV), Rome, Italy

Giuliana Panieri Institute of Polar Sciences, National Research Council, Venice, Italy

Giuseppe Puglisi National Institute of Geophysics and Volcanology (INGV), Rome, Italy

Vincenzo Romano National Institute of Geophysics and Volcanology (INGV), Rome, Italy

Mauro Sclavo Institute of Polar Sciences, National Research Council, Venice, Italy

Simona Scollo National Institute of Geophysics and Volcanology (INGV), Rome, Italy

Lijun Shang Biological Security Research Center, School of Human Sciences, London Metropolitan Universities, London, UK

Cristiano Tolomei National Institute of Geophysics and Volcanology (INGV), Rome, Italy

Daniele Trippanera National Institute of Geophysics and Volcanology (INGV), Rome, Italy

List of Figures

List of Tables

Chapter 1
Natural and Anthropogenic Gradients

Carlo Doglioni

Abstract Any type of phenomena is generated by a gradient. Therefore, also natural and anthropogenic hazards are also controlled by physical or socio-economic gradients. There are three types of hazards affecting the Euro-Mediterranean Region: (1) natural risks such as earthquakes, tsunamis, volcanic eruptions, floodings, landslides, geochemical degassing, etc. generated by physical gradients in the solid Earth and the atmosphere; (2) anthropogenic gradients such as the CO_2 rise determined the ongoing global warming generating higher evaporation of the marine areas, implying stronger and more frequent storms and precipitations inland, as well as the expansion of desertic regions, plus, sea-level rise threatening coastal areas due to ice melting in the polar regions; (3) cultural and socio-economic gradients between the several and variegated populations living all around the Euro-Mediterranean area may be considered as historical richness, but are instead often used as the trigger for international disputes, due to the higher demographic growth, contemporaneous desertification, and lower income per capita in Africa with respect to Europe, determining gradients able to fueling migration of the population from south to north. To be ready to face natural and anthropogenic catastrophic events, we need to increase our knowledge of the Earth's mechanisms concerning what we know at present to forecast their evolution. For example, we should all together work to immediately fight climate change, governing and lowering such influence of human beings on natural equilibria. We must know gradients in order to govern them, if possible.

Keywords Gradients · Natural hazards · Anthropogenic hazards · Socio-economic disparities · Euro-Mediterranean Region

C. Doglioni (✉)
Department of Earth Sciences, La Sapienza University, Rome, Italy
e-mail: carlo.doglioni@ingv.it

© The Author(s), under exclusive license to Springer Nature Switzerland AG 2025

M. F. Buongiorno et al. (eds.), *The Challenge of Environmental Security in the Euro-Mediterranean Region*, NATO Science for Peace and Security Series C: Environmental Security, https://doi.org/10.1007/978-3-031-99646-7_1

1.1 Introduction

A gradient is an increase or decrease in the magnitude of a given parameter observed in passing from one point or moment to another (Fig. 1.1). The gradient is a vector with a magnitude that quantifies the total amount of variation of a field per unit change in distance or time. The greater the gradient, the higher the energy involved. There is no phenomenon on Earth which is not related to a gradient: pressure, temperature gradient, electrochemical gradient, but also socio-political gradients [12, 13, 28]. The Mediterranean area is full of gradients, geological gradients, geodetic velocity differences [11], energy accumulation along fault zones, types of rocks, climate, etc. Africa and Europe will join sooner or later in a million years. We have a lot of seismicity in Italy, in the Mediterranean, we are exposed to this risk, it is a matter of security because earthquakes can generate big disasters in European countries, especially in Italy, Greece, Turkey, Syria, Jordan, Israel, western Northen Africa, where we expect major earthquakes to occur. Crete recorded historical events of M> 8 with consequent tsunami.

The Euro-Mediterranean region, a geographical crossroads linking Europe, Africa, and Asia, and it is characterized by its remarkable biodiversity, rich cultural heritage, and complex geopolitical landscape. However, this vibrant area is increasingly vulnerable to a confluence of interacting hazards. From the seismic activity that shapes its landscapes, to the growing impacts of human-induced climate change and the socio-economic disparities that can fuel instability, the region faces a multifaceted challenge. This paper aims to dissect these interconnected threats, focusing on the three primary categories identified: natural risks generated by the Earth's physical gradients (Fig. 1.2), anthropogenic hazards driven by the escalating CO_2 rise, and cultural/socio-economic gradients that can unfortunately trigger international disputes.

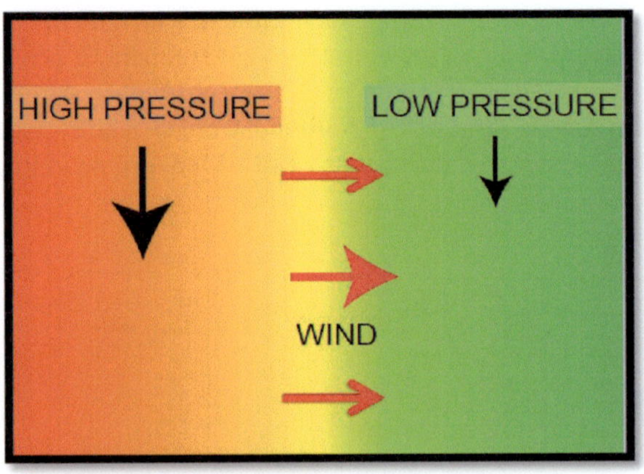

Fig. 1.1 A gradient is a variation of a parameter, and the greater the gradient, the stronger the energy that it generates

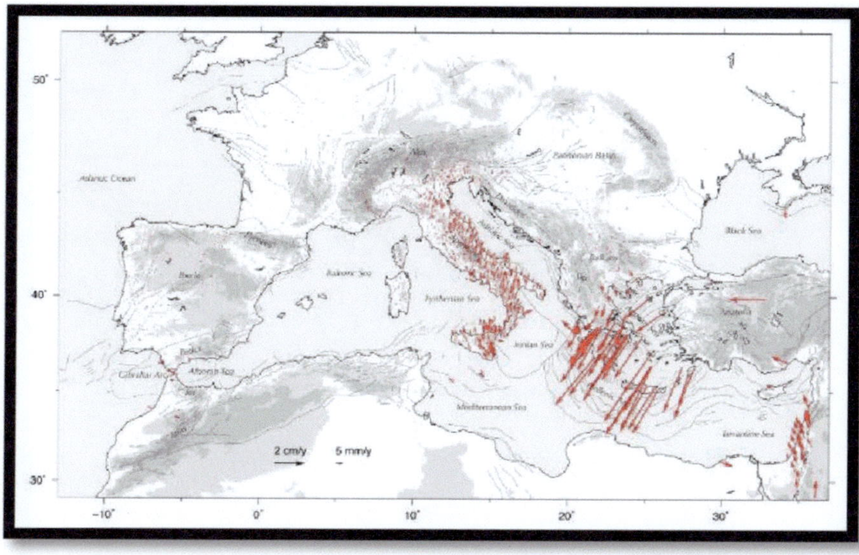

Fig. 1.2 Map of the horizontal GPS velocities (in the Eurasian-fixed frame) from the combined solution. Error ellipses are not shown here for clarity of the figure, after [11]. Notice the few data from Africa where the area is barely monitored. The velocity gradients determine plate motion and seismicity in the area

By exploring the nature and interplay of these hazards, this paper underscores the urgent need for enhanced scientific understanding and collaborative action to safeguard the future of the Euro-Mediterranean region.

1.2 Natural Risks

The Euro-Mediterranean basin (Fig. 1.3) is a geologically active zone [2, 16], positioned at the convergence of major tectonic plates (Africa and Eurasia) and several smaller plates (Corsica-Sardinia, Sicily, Adriatic, Aegean, Anatolia).

This dynamic setting makes it inherently susceptible to a range of natural hazards [7, 8]. Earthquakes [24], a recurring threat (Fig. 1.4), have historically caused widespread devastation, a stark reminder of the powerful forces shaping the Earth's crust beneath the region, although nowadays we still do not know what the engine of plate tectonics is.

The hazard due to tsunamis [5, 21, 25] can be triggered by submarine earthquakes or volcanic eruptions, subaerial and submarine landslides add another layer of risk to inland areas and coastal communities [6, 10]. The region also hosts several active volcanic areas, with eruptions posing dangers from lava flows and ashfall to the release of hazardous geochemical degassing [4, 20, 23]. Furthermore, the varied

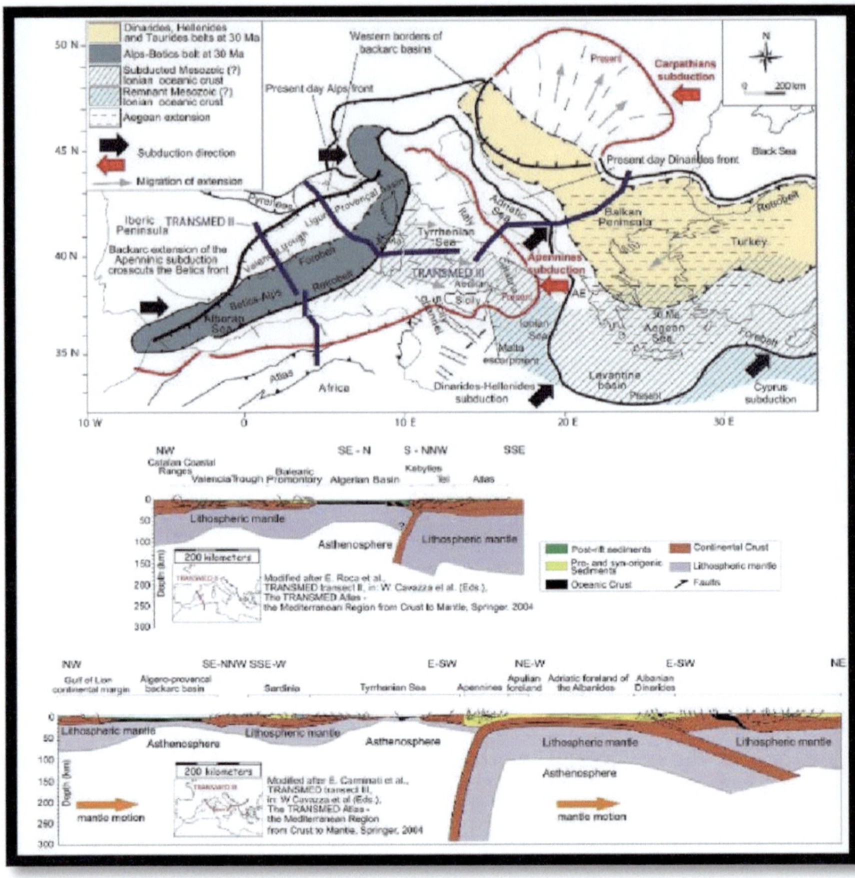

Fig. 1.3 Present geodynamic framework of the Mediterranean region. Four subduction zones are active in the Mediterranean area: the W-directed Apennines-Maghrebides; the W-directed Carpathians; the NE-directed Dinarides-Hellenides-Taurides; and the SE-directed Alps. Two lithospheric-scale cross-sections illustrate the mountain belts, and the basins associated with these subductions. AE, Apulian escarpment. Modified after [8]

topography and climatic conditions contribute to the prevalence of landslides, particularly in mountainous areas, and both riverine and coastal flooding events, exacerbated by intense rainfall and sea-level fluctuations. These natural hazards, intrinsic to the region's physical geography, demand continuous monitoring, risk assessment, and robust preparedness strategies.

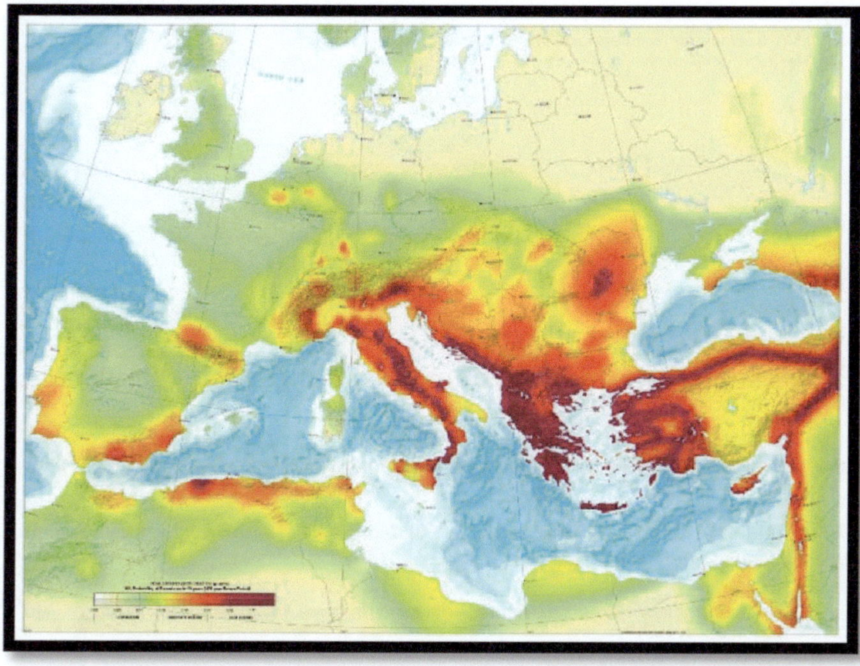

Fig. 1.4 Horizontal peak ground acceleration seismic hazard map representing stiff site conditions for an exceedance or occurrence rate of 10% within 50 years for the Mediterranean Region, after [15]

1.3 Anthropogenic Gradients

Superimposed on this backdrop of natural vulnerabilities are the increasingly severe anthropogenic hazards, primarily driven by the unprecedented rise in atmospheric carbon dioxide concentration. This surge in CO_2, largely attributable to human activities such as the burning of fossil fuels and deforestation, is the principal determinant of ongoing global warming (Fig. 1.5), with profound consequences for the Euro-Mediterranean [17].

The warming climate is intensifying the hydrological cycle, leading to higher evaporation rates over marine areas, which in turn fuels stronger and more frequent storms and precipitation inland, increasing the risk of severe flooding [1, 22]. Simultaneously, rising temperatures contribute to the expansion of desertic areas, particularly in the southern Mediterranean, threatening agricultural productivity and water resources. Critically, sea-level rise [27], primarily driven by the melting of polar ices, poses an existential threat to low-lying coastal areas, impacting infrastructure, ecosystems, and human settlements (Fig. 1.6).

These anthropogenic hazards are not merely environmental concerns; they have far-reaching socio-economic and geopolitical implications.

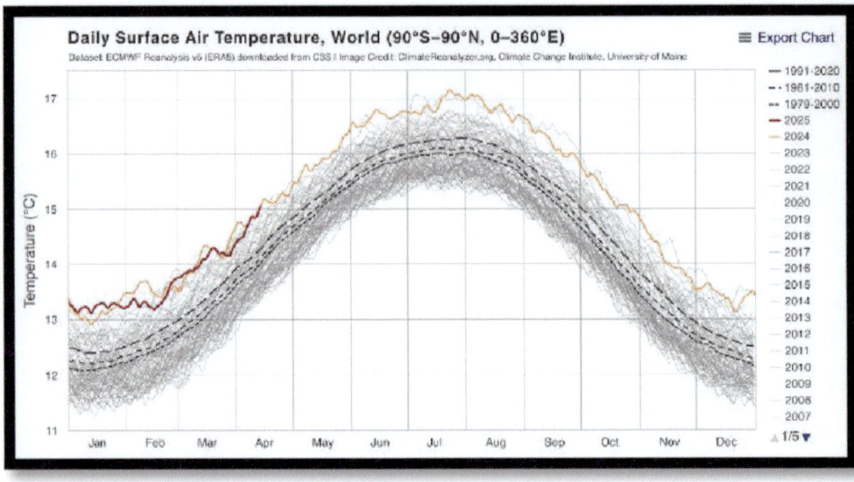

Fig. 1.5 World daily surface air temperature, World. Notice the clear increasing trend. Data from https://climatereanalyzer.org

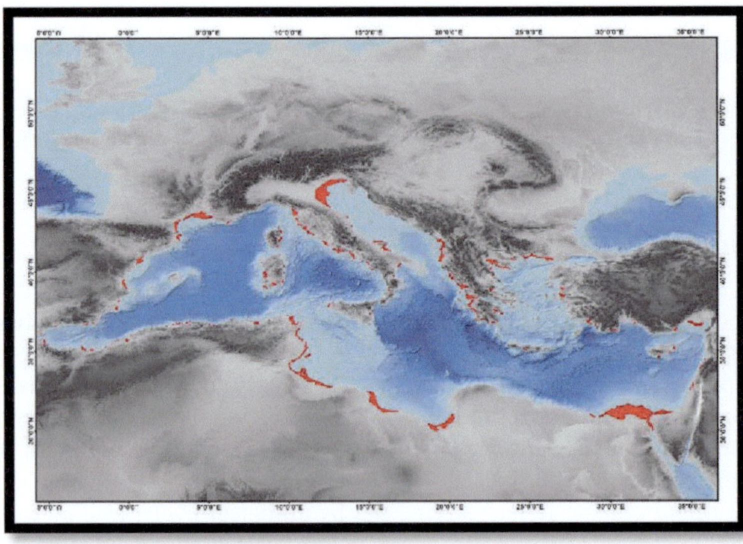

Fig. 1.6 The main coastal plains of the Mediterranean basin which are expected to be submerged by the sea-level rise due to global warming are underlined in red. The geospatial analysis highlights that the total area of the coastal plains of the Mediterranean region potentially prone to be flooded is about 38 529 km^2. The most affected countries are Egypt (12.879 km^2), Italy (10.060 km^2) and France (3.681 km^2). After Vecchio et al. [27]

1.4 Cultural and Socio-economic Gradients

The Euro-Mediterranean region boasts a rich tapestry of cultures, languages, and socio-economic systems, a legacy of its long and complex history [9, 14, 26]. While this diversity represents an invaluable asset, these cultural and socio-economic gradients can unfortunately become flashpoints for international disputes. Competition for scarce resources, such as water and arable land, can be intensified by environmental degradation. Disparities in economic development and social well-being across the region can fuel migration patterns and create tensions between nations. Historical grievances and differing political ideologies can further exacerbate these underlying socio-economic fault lines. Recognizing the potential for these gradients to act as triggers for conflict is crucial. Transforming these differences from sources of division into opportunities for collaboration, mutual understanding, and shared prosperity is a significant challenge but also a vital necessity for regional stability.

In rich countries, there is a high concentration of raw materials or oil deposits whose rents, however, are not the major consumers of these goods: these gradients determine migration flows, commodity markets, the value of currencies, and energetic resources. These gradients control the relationships between nations.

The "negative" gradient is generated for example between two populations that for cultural or economic reasons fail in producing a flow towards mutual cooperation. The diplomatic apparatus has precisely the task of dissolving this specific gradient to avoid a disastrous evolution in a situation of war. The cultural "negative" gradient is to be confronted with the difference in traditions, history, religion and language between communities or nations, but considerable wealth is needed to tutorize. The socio-economic gradient is demonstrated as a fundamental factor in the incidence of determined pathologies; where in general the nation richest in resources suffer from minor pathologies, due to higher per capita spending, in particular, there is a direct correlation between the nations that have greater availability of potable water and the respective life expectancy. In poor countries, the poorest people have lower access to even the most basic resources and this gradient entails an important tentative to improve the proper conditions of economic income. Depression post-partum or the development of diabetes are major diseases due to the high per capita income. There is an important sanitary gradient between rich and poor countries, up to the internal nations as a function of health status. The increase of the gradient of health status generates a higher risk for the health of citizens that, according to the World Health Organization, is the summit of all the risks, environmental imbalance, the lack of instruction, the quality and quantity of food available, the lack of work, habitations and internal poverty. Attention for the poorest and weakest diminishes the internal social gradient within a nation and tends to re-establish political and social tensions. However, there are still many gradients that differentiate salary administrators of industry and operations, a gradient that becomes embarrassing and generator of tensions social and comprehensibility union demands when this relationship surpasses a tolerable ratio. The demographic growth in Africa from 1950 to

2050 is foreseen at 73%. In Europe it is only 11%, a testimony of how notoriously the economic comfort brings a minor demographic growth (Fig. 1.7).

The coupling between overpopulation and poverty speeds up the migration flows from Africa to Europe, from poor to richer countries (Fig. 1.8).

The demographic evolution remains a fundamental issue in this first order analysis [3, 18, 19].

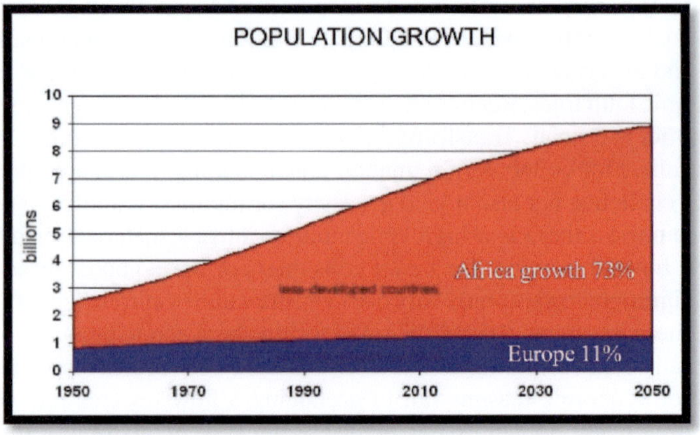

Fig. 1.7 Comparison of the demographic evolution of Africa and Europe. Data from United Nations, World Population Prospects. It is evident the European decline with respect to Africa

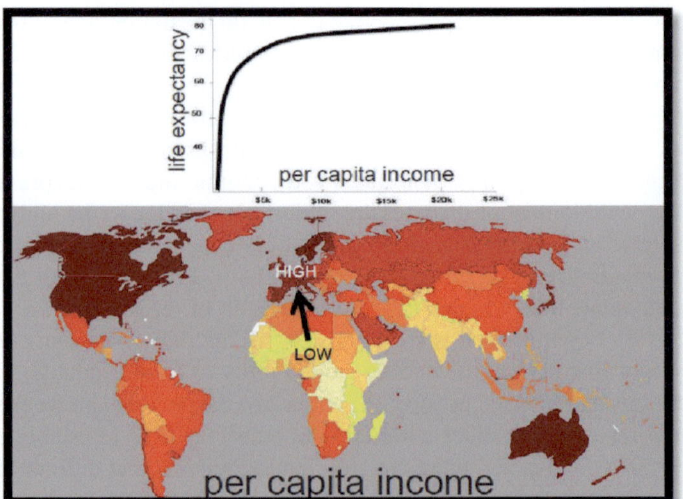

Fig. 1.8 The lower income per capita clearly controls the life expectance (upper panel) and triggers southern populations to migrate northward in areas with higher income (lower map), where global warming still allows widespread agriculture and better quality of life

1.5 The Interconnectedness of Hazards

It is crucial to recognize that these three categories of hazards are not isolated phenomena; rather, they are intricately interconnected and can amplify each other's impacts. For instance, climate change-induced extreme weather events can increase the likelihood and severity of natural disasters like floods and landslides, disproportionately affecting vulnerable socio-economic groups and potentially exacerbating existing inequalities. Environmental degradation driven by anthropogenic factors can further strain resources, intensifying competition and potentially contributing to social unrest and international tensions. Conversely, socio-economic instability can hinder effective responses to both natural and anthropogenic hazards, impeding mitigation and adaptation efforts. Understanding these complex interactions is paramount for developing holistic and effective risk management strategies.

1.6 The Urgent Need for Increased Knowledge and Collaborative Action

To effectively confront this triple threat, a significant increase in our understanding of the Earth's intricate mechanisms is paramount. This necessitates sustained investment in scientific research, advanced monitoring systems, and open data sharing across the Euro-Mediterranean region. Furthermore, all this emphasizes the immediate need to fight climate change by governing and lowering the influence of human beings on natural equilibria. This requires ambitious mitigation policies aimed at reducing greenhouse gas emissions and transitioning to sustainable energy systems. Concurrently, adaptation strategies are essential to build resilience to the unavoidable impacts of climate change. Crucially, addressing the socio-economic dimensions requires fostering dialogue, promoting equitable development, and strengthening regional cooperation to transform potential conflict drivers into avenues for shared progress and stability.

1.7 Conclusions

Science may be second lever with respect to peace, but science may help to bring peace. The Euro-Mediterranean region faces a complex interplay of hazards categorized as natural (earthquakes, tsunamis, volcanic eruptions, etc.), anthropogenic (driven by CO_2 rise and global warming), and cultural/socio-economic (gradients acting as triggers for disputes). This paper explores the distinct characteristics of each hazard category and, critically, their interconnectedness. It argues that the region's inherent natural vulnerabilities are being exacerbated by human-induced climate change, while existing socio-economic disparities can escalate tensions in the face of environmental and resource pressures. Addressing this triple threat necessitates a significant increase in our understanding of Earth's systems and, crucially, immediate and collaborative action to mitigate anthropogenic impacts and foster cooperation across diverse populations, transforming potential conflict drivers into opportunities for regional resilience.

The Euro-Mediterranean region stands at a critical juncture. It faces a complex and interconnected web of natural, anthropogenic, and socio-economic hazards that pose significant challenges to its environmental sustainability, economic prosperity, and social cohesion, being an extraordinary crossroad of different cultures and religions. While the region's inherent natural vulnerabilities are undeniable, the escalating impacts of human-induced climate change and the potential for socio-economic gradients to fuel instability demand urgent attention and concerted action. By prioritizing scientific understanding, implementing ambitious climate policies, and fostering genuine collaboration across diverse populations, the Euro-Mediterranean region can strive to transform these threats into opportunities for building a more resilient, sustainable, and peaceful future for all its inhabitants.

References

1. Allen M, Dube OP, Solecki W, Aragón-Durand F, Cramer W, Humphreys S, Kainuma M (2018) Special report: global warming of 1.5 C. Intergov Panel Clim Chang (IPCC) 677:393
2. Ambraseys N (2009) Earthquakes in the Mediterranean and Middle East: a multidisciplinary study of seismicity up to 1900. Cambridge University Press
3. Ambrosetti E (2020) Demographic challenges in the Mediterranean. MED 300–304
4. Beccaluva L, Bianchini G, Wilson BM eds (2007) Cenozoic volcanism in the Mediterranean area, vol 418. Geological Society of America
5. Billi A, Funiciello R, Minelli L, Faccenna C, Neri G, Orecchio B, Presti D (2008) On the cause of the 1908 Messina tsunami, southern Italy. Geophys Res Lett 35(6)
6. Camerlenghi A, Urgeles R, Fantoni L (2010) A database on submarine landslides of the Mediterranean Sea. In: Submarine mass movements and their consequences, pp 503–513
7. Carminati E, Lustrino M, Doglioni C (2012) Geodynamic evolution of the central and western Mediterranean: tectonics versus igneous petrology constraints. Tectonophysics 579:173–192. https://doi.org/10.1016/j.tecto.2012.01.026
8. Carminati E, Petricca P, Doglioni C (2021_ Mediterranean tectonics. In: Encyclopedia of geology, 2nd edn. Elsevier, pp 408–419. https://doi.org/10.1016/B978-0-08-102908-4.00010-21
9. Costa-i-Font J ed (2012) Europe and the Mediterranean economy. Routledge, Oxon
10. Crosta GB, Frattini PJNH (2003) Distributed modelling of shallow landslides triggered by intense rainfall. Nat Hazard 3(1/2):81–93
11. Devoti R, d'Agostino N, Serpelloni E, Pietrantonio G, Riguzzi F, Avallone A, Cavaliere A, Cheloni D, Cecere G, d'Ambrosio C, Falco L (2017) A combined velocity field of the Mediterranean region. Ann Geophys 60(2)
12. Doglioni C (2021) Gradienti della Terra e non solo. Altraparola. https://www.altraparolarivista.it/2021/03/13/gradienti-della-terra-e-non-solo-di-carlo-doglioni/
13. Doglioni C (2024) Gravitational and elastic energies stored in crustal volumes activate normal versus strike-slip and thrust seismogenic faults. Geosci Front. https://doi.org/10.1016/j.gsf.2024.101894
14. Gilmore DD (1982) Anthropology of the Mediterranean area. Annu Rev Anthropol 175–205

15. Grünthal G, Bosse C, Sellami S, Mayer-Rosa D, Giardini D (1999) Compilation of the GSHAP regional seismic hazard for Europe, Africa and the Middle East. Ann Geophys 42(6)
16. Grünthal G, Wahlström R (2012) The European-Mediterranean earthquake catalogue (EMEC) for the last millennium. J Seismolog 16:535–570
17. Kogan F (2023_ The IPCC reports on global warming and land changes. In: Remote sensing land surface changes: the 1981–2020 intensive global warming. Springer International Publishing, Cham, pp 67–79
18. Livi Bacci M (2001) Demographic shocks: the view from history. Popolazione e storia 2(2):93–114
19. Livi Bacci M (2017) A concise history of world population. Wiley
20. Lustrino M (2000) Volcanic activity from the Neogene to the present evolution of the Western Mediterranean area. a review. Ofioliti 25(2):87–101
21. Maramai A, Brizuela B, Graziani L (2014) The euro-mediterranean tsunami catalogue. Ann Geophys 57(4):S0435–S0435
22. Masson-Delmotte V, Zhai P, Pörtner HO, Roberts D, Skea J, Shukla PR (2022) Global warming of 1.5 C: IPCC special report on impacts of global warming of 1.5 C above pre-industrial levels in context of strengthening response to climate change, sustainable development, and efforts to eradicate poverty. Cambridge University Press
23. Peccerillo A (2005) Plio-quaternary volcanism in Italy, vol 365. Springer, New York, Berlin, Heidelberg
24. Petricca P, Carminati E, Doglioni C (2022) Estimation of the maximum earthquakes magnitude based on potential brittle volume and strain rate: the Italy test case. Tectonophysics. https://doi.org/10.1016/j.tecto.2022.229405
25. Selva J, Lorito S, Volpe M, Romano F, Tonini R, Perfetti P, Bernardi F, Taroni M, Scala A, Babeyko A, Løvholt F, Gibbons SJ, Macías J, Castro MJ, González-Vida JM, Sánchez-Linares C, Bayraktar HB, Basili R, Maesano FE, Tiberti MM, Mele F, Piatanesi A, Amato A (2021) Probabilistic tsunami forecasting for early warning. Nat Commun 12(1):5677
26. Sherratt S (2003) The Mediterranean economy: 'Globalization' at the end of the second millennium BCE. In: Symbiosis, symbolism and the power of the past, pp 37–62
27. Vecchio A, Anzidei M, Serpelloni E (2023) Sea level rise projections up to 2150 in the northern Mediterranean coasts. Environ Res Lett 19:014050. https://doi.org/10.1088/1748-9326/ad127e
28. Zaccagnino D, Doglioni C (2022) Earth's gradients as the engine of plate tectonics and earthquakes. La Rivista del Nuovo Cimento. https://doi.org/10.1007/s40766-022-00038-x

Chapter 2
Environmental Security and Climate Change: Strategic Scenarios and Crisis Factors

Enrico Casini and Federico Deiana

Abstract Security threats arising from the environment and climate change are not limited to asymmetric conflicts and non-state actors but are also increasingly influencing conventional military operations. Global warming, rising sea levels, and extreme weather events are forcing changes in defense strategies, impacting military infrastructure, operations, and personnel preparedness. Armed forces will need to adapt to new climatic conditions, facing challenges such as increased air turbulence or changes in maritime routes. Additionally, environmental changes are reshaping geopolitical geography, creating new opportunities and conflicts over control of natural resources, such as water and minerals essential for the energy transition. Coastal cities, vulnerable to natural disasters, pose a threat to internal security, as critical infrastructures can be damaged, as evidenced by events like Hurricane Katrina. Finally, climate change is becoming a central geopolitical factor, with potential rivalries among global powers to exploit transforming territories and resources, such as the Arctic. Managing these challenges will determine future security and stability, requiring international cooperation and investment in resilience.

Keywords Enviromental security · Climate change · Threat multiplier · Defense · Geopolitical competition

2.1 The Expansion of the Concept of Security

The term "Security" is commonly used to refer to the condition "that makes one feel free from danger" or that "provides the possibility to prevent, eliminate, or mitigate damage, risks, and difficulties" [1]. Throughout human history, security has always been one of the key conditions upon which the stability and solidity of societies have been based. Therefore, from ancient times, it has also been one of the main objectives

E. Casini (✉) · F. Deiana
Med-Or Italian Foundation, Rome, Italy
e-mail: enrico.casini@leonardo.com

© The Author(s), under exclusive license to Springer Nature Switzerland AG 2025

M. F. Buongiorno et al. (eds.), *The Challenge of Environmental Security in the Euro-Mediterranean Region*, NATO Science for Peace and Security Series C: Environmental Security, https://doi.org/10.1007/978-3-031-99646-7_2

that governments or rulers, both in autocratic or absolutist regimes and in democratic ones, have sought to guarantee, both to protect their own power structure and, more generally, to ensure social peace and well-being.

Historically, the main threat to the security of a state, a defined territory, or its inhabitants has been represented by war and armed conflict, and more generally by the use of force; so much so that the first organized statehoods were formed around the exercise of this monopoly, continuing through to the emergence of modern states from the fifteenth century onward. In fact, throughout human history, the security of social organizations and populations has often been endangered by natural and environmental factors, as well as health-related events. Cataclysms such as floods, earthquakes, or volcanic eruptions, as well as droughts or sudden temperature changes, and even epidemics and pandemics, have often contributed to the sociopolitical changes that emerged in certain historical periods. It is no coincidence that recent historiographical studies have attempted to assess how significant the impact of epidemics, climate change, and natural disasters has been compromising not only the economic-social and political stability of affected regions but also their defense capabilities. For example, the epidemics that struck the Roman Empire in the second and third centuries A.D. may have weakened the Empire's defensive capacity along its borders, just as the Justinian Plague, combined with a phase of cooling, may have contributed to the general, even military, weakening of the Byzantine Empire [2].

Over time, through the complex course of various historical eras, the factors representing threats to security have increased, as well as transformed. In addition to the use of organized force and violence, other types of threats of various kinds have emerged, some originating from human actions, others from natural or environmental causes.

After the end of the Second World War, there was an attempt to curb organized violence and wars through forms of international regulation and law, with the United Nations project being the most prominent expression of this effort. However, this did not prevent the outbreak of conflicts across the globe, nor the emergence of threats to the security of communities and states. Terrorism, asymmetric warfare, insurrections, local conflicts, and identity-based struggles were some of the forms in which violence manifested itself in an irregular and unconventional manner, alongside more traditional wars between states. However, after Hiroshima, and with the beginning of the Cold War, from the second half of the twentieth century onward, the world was also introduced to a new threat: nuclear war. This new power of deterrence radically changed the strategic paradigm of international security and the very way of conceiving large-scale conflicts [3]. More recently, with the information revolution and the dawn of what we might call the "new digital age," we are entering a new era of security: the classic dimensions in which threats and conflicts were measured are now extending beyond the physical borders of the known world, into both the virtual and spatial dimensions.

Wars now seem to have no end, neither in time nor in spatial boundaries. This transformation has made the realm of security increasingly multifaceted and global, conditioned not only by traditional threats but also by other global factors. This is due to historical processes, alongside the technological revolution, that have made the world interconnected and interdependent. Indeed, since 1945, we have witnessed the emergence of a growing interdependence among actors in the international system, which, on the one hand, led to the initiation of globalization, but also influenced the sphere of security, expanding its boundaries immeasurably. Economies and markets, supply chains, logistics, technologies, political systems, and even the medical sector have become increasingly interdependent. This ongoing process of changing integration, further accelerated by the end of the division of the world into two blocks, has meant that events occurring in a particular national or regional context could have enormous consequences on the other side of the planet. The decrease in oil production in the Middle East during the first crisis in 1973, for example, led to an energy shock worldwide, starting with Europe and North America; just as the financial bubble burst in Southeast Asia compromised the competitiveness of South American economies. Furthermore, with such a regime of interdependence, it is not surprising that a food crisis or the spread of a virus could easily transform into a continental famine or a pandemic, with severe repercussions across many aspects of life [4].

All of this has had a significant impact on security. The deep interconnection of the various sectors that directly or indirectly affect it, and the extreme ease with which the chain of events, even from afar, can have observable consequences in the lives of each of us, has led to a considerable expansion of the concept and the major sectors of Security in the second half of the twentieth century [5]. This expansion in an increasingly "global" sense of the very concept of security has gone hand in hand with the progressive expansion of the concept of risk itself. Paraphrasing Ulrich Beck [6] and his studies, humanity now lives in a global risk society, in which, as he states at the beginning of his most famous work, "In modernity, the social production of wealth goes hand in hand with the social production of risks" [7]. This multiplication of global risk factors and threats now also includes the environment and climate, particularly in this historical period. Even today, as in the past, within the broad context of contemporary International Relations, the military domain remains dominant in security and defense logics. However, it is increasingly accompanied by economic, energy, health, food, and cultural components, along with numerous other subdomains of growing importance. With the emergence of the new complexity that characterizes the dimension of security in the contemporary age, and the increasing evidence of the importance of natural hazards in political and scientific debates, new elements have begun to gain ground in security studies since the 1970s: in particular, Climate Change and the risks for the Environment [8].

2.2 Environmental Security

In the broader context of environmental security and the threats arising from catastrophic natural events and environmental factors, climate and its changes have become increasingly significant. The more evident and radical the changes in environmental and climatic balances have become, as observed in recent years, the more these issues have gained importance and drawn increased attention. Indeed, since the topic of climate change has assumed an ever more prominent position within natural sciences, social sciences have also begun to study its implications, particularly from economic and political perspectives. With the growing awareness of the scope of the issues and consequences related to environmental security and climate change, it has become increasingly clear that the events associated with them may have repercussions not only at an exclusively ecological and environmental level but also across all other facets of human society. In particular, climate change will be one of the potential threats to security on which our attention will be focused, due to its wide-ranging, cross-cutting nature.

Climate Change refers to the "long-term variation in the Earth's average temperature and weather conditions," the most obvious manifestation of which is global warming, with consequences extending across a significant number of sectors and fields [9].

The higher average temperatures recorded in recent years are, for example, accelerating the melting of polar and mountain glaciers in a process that appears increasingly rapid [10], leading to rising sea levels, changes in the chemical-physical composition of ocean water, and alterations in temperature and underwater currents. The increasingly frequent extreme weather events constitute another concrete manifestation of climate change, as well as saline intrusion into freshwater bodies, numerous episodes of drought, desertification, and soil erosion.

The first studies on Climate Security focused on analyzing a potential positive correlation between climate change and insecurity in fragile contexts around the world. The goal was to demonstrate the existence of a Climate Change-Security Nexus [11], to understand if certain climatic conditions could worsen the economic, political, and social situation of some communities, to the point of increasing insecurity in a given area and causing wars or other destabilizing phenomena.

Over time, it has been demonstrated that each of these climatic factors, in turn, impacts the security of people and states. Desertification and drought threaten the availability of water resources for domestic use, but also for economic activities such as electricity generation, industrial activities, and especially agriculture and livestock. Floods and inundations, originating from increasingly unpredictable and catastrophic weather events, can damage infrastructure and cities, causing enormous losses. At the same time, climate alterations can cause severe damage to ecosystems, often resulting in a loss of biodiversity and—especially in the case of oceans—drastic reductions in food resources available for human consumption. Furthermore, higher temperatures can facilitate the development of large-scale wildfires, and from a health perspective, the extension of hot or cold seasons is often associated with the easier spread of diseases and invasive species.

2.3 Climate Change as a "Threat Multiplier"

Over the last twenty years, studies on Climate Security have further progressed. Thanks to the most recent analyses of the pervasive effects of climate change, along with the publication of numerous scientific studies supported by field data, it has been understood that, rather than investigating whether there is a link between the two elements—climate and security—now taken for granted, it is more crucial to focus on how and when climatic factors can influence the stability of countries and the lives of the citizens living in them [12]. This essential shift also happened due to the progressive worsening and intensification of the disastrous events described earlier, which have become increasingly frequent compared to the past, with more severe consequences for the affected communities. Recently, there have been multiple examples of such events, from Spain to the United States to Italy. In addition to the risks posed by extreme weather events, such as floods, or long-term processes like desertification, it has been realized that climate factors can have a systemic impact on a series of security-related challenges, with a potential domino effect. In academic literature on the subject, the concept of the "Threat Multiplier" has emerged: climate change is now seen as an independent variable that can cause or exacerbate already critical situations, characterized by fragility, conflict, and competition for scarce resources [13].

Consider the case of many developing countries. In addition to coinciding largely with the geographical areas of the planet most exposed to climate change, these countries often see the majority of their population engaged in agriculture and livestock—the two economic sectors most affected by the seasons and, therefore, dependent on climate factors. These areas are home to various population groups still living in widespread poverty, with a lack of basic services, outdated technologies, and the constant risk of sudden food crises, which, as seen recently with the Russian-Ukrainian wheat issue, can also stem from exogenous factors, in addition to endogenous ones. In such contexts, events like floods or droughts can turn into real catastrophes, where the absence of updated tools and emergency plans to respond to natural disasters can further complicate the situation. In 2022 alone, for example, the prolonged worst drought in 40 years in the Horn of Africa caused crop yields to be between 50 and 95% lower than the previous years' average, and over 8 million livestock deaths. However, beyond the specific data regarding the disastrous year of 2022, it is the consequences of the following years that truly explain the magnitude of the phenomenon in question: estimates for 2024 suggest that over 25% of the population in the region has experienced high levels of acute food insecurity due to prolonged drought. Currently, around 50 million people in East Africa need humanitarian assistance, and, as if the situation were not already alarming enough, the World Health Organization has recently stated that the food crisis could turn into a health catastrophe: in addition to severe malnutrition, there is a high risk of outbreaks of diseases such as cholera, malaria, and dengue fever [14].

As for long-term processes, the threat-multiplier effect of climate change is even greater. When not dealing with a sudden event but with the gradual deterioration of conditions in a region or specific area, broader potential security consequences can be observed, especially in the political and socio-economic context. Staying in Africa, consider the case of the Sahel. Since the second half of the twentieth century, global warming has led to an increase in average annual temperatures in the region by + 0.7 to + 0.9 °C. During this period, already scarce rainfall has further contracted, resulting in extended drought periods, intense rain episodes concentrated in a few days and relatively small areas. Coupled with the rapid demographic growth characterizing the region, this is creating an increasingly fierce competition for natural resources—primarily freshwater—in which pre-existing ethnic tensions between the various groups living in the area can easily be exacerbated [15].

In this regard, in the areas most severely affected by climate change, it is not difficult to find ethnic or religious minority groups compared to the rest of the country, complaining of an unfavorable distribution of natural resources and land allocated to them by the central government. This is the case with some minorities in Sub-Saharan Africa, as well as in Central and Southern Asia. In such contexts, a further reduction in available water or a livestock die-off due to parasites or diseases that are more conducive to the constantly rising temperatures can lead to protests, clashes, or even full-scale civil conflicts. Of course, in all these cases, climate should not be seen as the key factor in explaining such social phenomena. It is, in fact, a threat multiplier that exacerbates already existing issues.

A similar discourse can be made regarding migration issues. In contexts of widespread poverty or where there are persecutions and wars, today a growing number of people, compared to the past, are choosing to migrate from their home regions to seek better living conditions elsewhere. One increasingly significant fact in this regard is the phenomenon of climate migrants, a broad category of individuals that includes both international migrants and especially internally displaced people (IDPs) —that is, those forced to move but remaining within their national territory. United by the triggering cause of their displacement, climate change, this category of migrants grows every year worldwide: according to the latest data from the International Organization for Migration (IOM), since 2008 approximately 376 million people have been forced to migrate due to climate factors, 32 million of whom were displaced in 2022 alone [16]. Projections for the future even suggest that in the worst-case scenario, 1 billion people could become climate migrants by 2050, moving either internally within their countries or across borders, permanently or temporarily [17]. If this were to occur, the repercussions of such movements would be incalculable from a social, economic, and security standpoint, both for the destination countries—which would need to absorb such numbers into their national contexts—and especially for the areas of origin of the climate migrants, which would see further depletion of the productive fabric due to the progressive and continuous departure of individuals from the region.

2.4 Environment, Climate and Asymmetric Conflicts

In addition to migration, another crucial aspect to consider—concerning both natural and environmental events, and more specifically the impact of climate change—relates to the domain of defense and security in a narrow sense, particularly in terms of armed conflict and organized violence.

Indeed, in recent years, the growing influence of environmental factors and climate change on criminal activity, terrorism, non-state violence, and asymmetric warfare has been widely documented. These environmental and climatic elements can act as threat multipliers, especially in poorer or politically unstable countries, where their negative impact tends to be more severe.

A global analysis reveals a clear correlation between regions most exposed to the severe consequences of climate change—often also marked by political and institutional instability—and a higher prevalence of conflicts. These are frequently of an insurgent, internal, or asymmetric nature, and involve terrorist, irregular, or criminal organizations.

Such regions, ranging from Sub-Saharan Africa to South America and South Asia, are commonly characterized by rapid population growth. At the same time, they often coincide with the areas most affected by serious environmental and climatic challenges. In many of these countries, the reduction of arable land or scarcity of water and other natural resources—whether due to climate change or unsustainable land exploitation—has endangered lawful economic activities and the survival of entire communities. This situation has facilitated the recruitment efforts of criminal organizations, increasing the incidence of smuggling and human trafficking. Moreover, it has also promoted the enlistment of individuals into terrorist or insurgent groups in their campaigns against state or local institutions.

Economic grievances have at times intertwined with tribal, religious, or ethnic rivalries, exacerbating pre-existing tensions and conflicts. From this perspective, the Sahel region—particularly the Lake Chad Basin—is emblematic, as are numerous Middle Eastern countries such as Iraq and Yemen.

In some of the world's most fragile and impoverished countries—especially those afflicted by climate change and environmental degradation—criminal and terrorist activities have also flourished for a specific socioeconomic reason: the partial or total lack of viable socioeconomic alternatives. As highlighted by recent theories in behavioral and development economics [18], in such contexts, the opportunity cost of engaging in illegal activities is relatively low. Where economic marginalization, widespread poverty, and limited social mobility prevail, a growing number of individuals may find it more advantageous to turn to criminal enterprises than to engage in less lucrative legal economic sectors. Under these conditions, daily life becomes increasingly difficult and precarious, prompting more individuals to turn to criminal activities or to affiliate with irregular, including terrorist, militias to advance their causes or express their grievances. Unsurprisingly, the regions most at risk of becoming new hubs for international terrorism are precisely those experiencing the highest levels of inequality and poverty, alongside acute environmental and climatic crises. These include Afghanistan, Sub-Saharan Africa, and the Red Sea region—three areas characterized by weak state control and, in some cases, the marginalization or even persecution of minority groups.

2.5 Strategic Implications for the Defense Sector

Security threats stemming from environmental and climatic factors, however, do not concern solely the activities of non-state actors or the realm of potential asymmetric threats and conflicts. Beyond the typical actors engaged in insurgent and asymmetric warfare—ranging from guerrilla groups to jihadist organizations—these factors are also increasingly affecting the more conventional dimensions of defense and security sectors. Their repercussions are already evident in certain operational theatres where regular armed forces are deployed, suggesting that future scenarios will be increasingly shaped by environmental and climate-related variables. The growing significance of these issues is such that they are beginning to generate strategic-level implications which, over time, will need to be taken into account both in terms of future operational planning and broader geopolitical considerations. As environmental and climatic conditions evolve and become more extreme, their influence on military operations, missions, and infrastructure will intensify—not only with respect to overseas engagements, but also concerning national defense and homeland security.

History is full of examples where sudden and unexpected changes in environmental and climatic conditions influenced the course of battles or entire military campaigns. Think of the classic case of Napoleon's campaign in Russia, or the torrential rains that affected the Battle of Waterloo to Napoleon's disadvantage. Looking at the state of many actual or potential war theaters, the threats linked to climatic conditions that may occur and the possibility of sudden catastrophic weather events foreshadow developments that cannot be ignored, both regarding military preparedness and the equipment and tools they rely on. Rising temperatures, water scarcity, worsening desertification in some regions, and the risks linked to extreme weather events will require an adjustment in all three levels (operational, tactical, and strategic) related to military missions. Over time, it will become necessary to adapt certain resources, both individual and collective, as well as combat methods, to the new environmental and climatic conditions.

On the other hand, climate change is also making substantial modifications to the physical conditions of numerous operational theaters, forcing armed forces to increasingly consider environmental factors when it comes to defending their territory and securing their borders. For example, the effects of rising sea levels pose risks to the security of various coastal infrastructures, both military, such as bases, and civilian strategic ones, such as ports and oil and gas extraction platforms, in addition to threatening coastal communities and cities. Rising sea levels are, in fact, linked to the phenomenon of glacial melt, which, along with rising water temperatures, could cause other types of transformations in ocean waters, impacting military operations, such as submarine activities [19]. Finally, there is also the significant aspect of the environmental impact of military activities, which affects both the operations on the field, such as combat, but also training, logistics, and the necessary infrastructure for their functioning.

It is therefore evident that the more radical and widespread the effects of climate change become, the more they could lead to substantial transformations in the classic dimension of security and defense, in addition to continuing to influence the spread of violence and asymmetric, unconventional conflicts.

Therefore, it is not surprising that some of the most advanced armed forces and defense systems around the world have long been considering the implications of climate change on military fronts. The impact that climate change-related events can have, and will have in the coming years, on specific operational theaters or numerous geostrategic scenarios has forced an increasingly in-depth strategic and political reflection to consider possible consequences and solutions to adopt.

NATO, for example, has long included climate change in its Strategic Concept, anticipating that in the future the effects of climate change could affect the security of military bases and critical infrastructures. At the same time, it will be necessary to operate in increasingly extreme weather conditions while maintaining the same level of precision, efficiency, and rapid intervention [20]. Ocean acidification may require more frequent maintenance of military ships, as many studies highlight a positive correlation between sea temperature changes and alterations in underwater sound propagation speed, with significant implications for conducting submarine operations [21]. Aerial missions may face more frequent episodes of severe turbulence, and extensive weather disturbances could lengthen the time of operations or even hinder the completion of required tasks. But think, too, of the impact that increasingly rapid and radical climate changes will have in strategically important areas like the Arctic and North Atlantic, where military activities have increased in recent year. In the future these developments, coupled with the possibility of rising geopolitical competition among global powers, could increase even further.

2.6 The Impact on the Internal Security of Countries

As seen, the security risks arising from climate change are not limited to poor countries or those most exposed to the consequences of global warming. The effects of the ongoing changes, which have drastically accelerated in recent decades, are affecting all regions [22]. For this reason, the current regime of interdependence makes every crisis, even if localized, potentially disastrous in geographically distant areas. Consequently, it is now clear that the strategic dimension of climate change also extends to national security.

At the level of states, in what is referred to as "homeland security," there is no sector of social life or the economy that is not affected by the risks associated with climate change. This makes it a particularly relevant topic when it comes to protecting strategic interests, threatened by their impact, or the security of communities and individuals. From critical infrastructures, often located in coastal areas, to industrial production, supply chains, logistics, urban environments, and densely populated areas. For instance, consider how many cities worldwide are currently threatened by the underestimated risk of rising sea levels due to the melting of polar ice. Or how urban areas are often those most severely affected by extreme weather events.

According to recent studies, by 2050, over 70% of the world's population will live in cities, with a large portion residing in cities located along coastal zones. These will mostly be medium to large urban agglomerations near coasts, along with approximately 65 megacities with over 10 million inhabitants [23]. With the intensification of extreme weather events and rising sea levels, the high population density in urban centers could amplify the effects of climate-related natural hazards. Recent examples abound: from the flood in Valencia, Spain, to the storm in Dubai, UAE, to Hurricane Milton in the United States and the landslides and flooding in Italy in recent years. In all these cases, early warning and prevention systems failed to provide sufficient protection against the risks and improve the security of the affected populations. The same applies to critical infrastructures damaged by such catastrophic events, as seen in September 2023 in Libya. The Cyrenaica area was hit by a tropical cyclone that destroyed the region's main roads, water, and electricity infrastructures. The toll was devastating: over 11,000 dead and around 9,000 missing, mostly due to the collapse of these infrastructures and the slow response from authorities to restore their functionality. Similarly, in 2005, Hurricane Katrina's destructive impact on American oil and gas infrastructure caused a global energy shock [24]. The disruption of production due to the destruction of extraction sites led to a sharp spike in oil and gas prices for several weeks, forcing oil-importing countries to resort to strategic reserves—similar to what happened after the outbreak of the First Gulf War or, more recently, after the large-scale invasion of Ukraine by Russia.

These risks, as shown by some of the examples reported, also affect the Mediterranean region, one of the most exposed to all major climate and environmental threats: desertification, wildfires, catastrophic storms, and drought. In this region, climate change can jeopardize many critical infrastructures on which the economy and stability of many countries depend, and more broadly compromise the strategic interests of regional actors.

As highlighted in recent years, the Mediterranean is one of the planet's areas most exposed to the risks of rising sea temperatures, which can contribute to the formation of extreme weather events like those affecting numerous coastal countries. It is also a region where, on the one hand, rising temperatures may lead to soil desiccation, extreme drought, wildfires, resulting in food crises, famine, significant economic damage. On the other hand, the fragility of many southern Mediterranean countries could be severely tested by the impact of "wild" climate, leading to increased insecurity and instability that threatens the already fragile socio-political balance in these countries, with potential implications for European security.

This combination of events and risks linked to climate change will require significant economic investments in many sectors in the coming years to address them. Building a more resilient system capable of reacting to these threats has become indispensable, not only for wealthier countries but also for the most fragile and poorer nations. This also requires international cooperation and collaboration to effectively address these challenges. A significant role can be played, as is already happening today, through the use of modern technologies and scientific research to find solutions for the prevention, management, or mitigation of their impact. Moreover, raising public awareness about the existing risks and ensuring greater sensitivity and preparedness among individuals will be increasingly useful.

2.7 Climate and Environment as Factors of Geopolitical Competition

Due to its growing strategic significance, the climate factor is gaining importance in the international political sphere, with a range of potential geopolitical repercussions. In the context of growing competition between world powers, amidst an international framework where political and economic balances are being redefined, climate could become an additional factor of instability and rivalry.

Many of the examples and cases mentioned above fall within the broad domain of international politics and have significant geopolitical implications. However, what we wish to examine in conclusion is the possibility that climate change, and environmental factors in general, could become a significant factor of geopolitical competition, in addition to being a source of risk.

In particular, it is interesting to consider two aspects: the first concerns the impact of climate change on ecosystems, natural environments, and natural resources; the second pertains to its ability to affect territories and regions on the planet by altering their natural conditions, and as a result, producing transformations in human geography.

Climate change, particularly the increase in atmospheric temperatures and the resulting changes—from reduced rainfall and soil aridification to transformations in the natural cycle of the seasons—along with the broad range of consequences these phenomena produce, initially leads to a reduction in essential and fundamental natural resources. Water and arable land, as well as the extinction of animal and plant species critical to the ecosystems and ecology of certain regions of the planet, are at stake. The reduction or potential disappearance of these resources in certain areas of the planet due to climate change could trigger a geopolitical and economic domino effect, sparking internal crises and tensions within the involved state organizations, as well as external effects, through tensions, conflicts, and competition for control over increasingly scarce or at-risk resources due to worsening climate conditions or environmental disasters. Indeed, this could lead to a significant shift in the global balance of power due to differing levels of access to increasingly precious resources. In the coming years, the rarer certain fundamental resources become—both for survival and for supporting the levels of economic development necessary to ensure well-being and stability for national communities—the more conflict over their control could intensify among states.

This could concern a resource that is becoming increasingly necessary for life, such as drinking water, which may potentially become scarcer in the future. A prime example of geopolitical competition related to this is the dispute over the waters of the Nile, as evidenced by the construction of the GERD dam and the resulting tensions between Egypt, Sudan, and Ethiopia.

In the future, countries rich in natural resources could see their international standing grow. In a future world that is hotter and has fewer available water resources, territories with abundant freshwater resources, for example, may leverage this advantage over drier regions. This can be referred to as "water weaponization," as well as "food weaponization" when the "blue gold" is used—or not used—to produce food resources for countries in greater need [25]. This logic, unfortunately, may apply to all resources whose distribution on the planet is in some way altered by climate change.

Moreover, as is already being observed on the international stage, the digital revolution and, in particular, the ongoing energy transition aimed at reducing the impact of fossil fuels on the environment and mitigating the risks associated with climate change, will drive a differentiation in energy sources through investments in renewables, nuclear power, and the electrification of various transportation sectors. This will lead to an exponential increase in the race for the acquisition of minerals essential for supporting digitalization and the energy transition. From uranium for nuclear power plants to lithium for electric batteries, and all minerals required for the production of microchips, the need to reduce CO2 emissions will drive an indispensable transformation of our societies, both technologically and energetically, already sparking a strategic confrontation between major world powers over dominance in emerging technologies, crucial for driving these processes, as well as control over materials and minerals essential for their development and use.

In addition, another aspect that could make climate change a factor of geopolitical competition is related to its effects on the geography of the planet. Indeed, it is not far-fetched to imagine that, due to climate change and the corresponding environmental and ecological transformations, new strategic scenarios could emerge in various regions of the globe. This may involve territories that become uninhabitable due to climatic changes, such as the more arid and hotter areas of the planet, but also those that could acquire more favorable conditions for human settlement and exploitation, such as areas near the polar circles. In both cases, significant changes could occur in human geography. The first, and perhaps most relevant, example concerns the Arctic and its surrounding areas, which, with the accelerating process of ice melt, is becoming an increasingly attractive region from an economic, commercial, logistical, and mining perspective [26]. In the coming years, as global temperatures rise, this polar region could become more habitable, potentially witnessing the establishment of new human settlements. However, the North Pole is also becoming the subject of an increasing militarization process, aimed at controlling its wealth and enhancing deterrence. Just as the Arctic regions may transform due to rising temperatures and the northern seas become more navigable, leading to profound changes in global transportation, it is also worth considering how some countries may view the impact of climate change on their territories as a positive factor, resulting from the transformation of previously uninhabitable or difficult-to-access areas into regions now open to colonization. A particularly emblematic case in this regard is Russia. In Russia, many mineral and energy resources are located in the country's colder regions. Additionally, the northern coasts are frozen for much of the year. Rising temperatures could make these regions more accessible and open up scenarios for exploiting territories, resources, and marine waters previously considered difficult or impossible to access. This holds true for Russia, but it could also apply to other countries, where climate change, by altering the geography of their territories, could represent an opportunity for development and a competitive advantage in the geopolitical race. What could be a problem for some may become an interesting opportunity for others to exploit, both geopolitically and economically. In this case, while climate change will undoubtedly have severe global environmental consequences, which will have negative repercussions for all in the medium to long term, some countries could opportunistically exploit it in the short term to gain an advantage in the global competition for geopolitical dominance. After all, geopolitics—understood as the manifestation of politics through geography—has always been driven by competition for resources or territorial control. The effects of climate change, due to their scale and the breadth of interests they affect and threaten, will increasingly become a central issue for both state security and international projection. No country will be able to ignore the need to secure access to primary goods, whose availability is at risk due to climate change.

2.8 An Uncertain Future

The future scenarios are highly uncertain, and it is impossible to predict today the outcomes of the ongoing environmental and climatic processes. However, it is clear that these processes are producing increasingly radical and swift changes that should not be underestimated. A hotter and more populated world may suddenly find itself thirstier and hungrier, with a reduced availability of arable land, and with all the consequences that such events could entail. In our future, entire areas of the globe could undergo profound transformations due to rising temperatures and their direct and indirect consequences. The fight against the most extreme forms of disaster produced by climate change could represent one of the most challenging and complex tests for populations and future governments. Millions of people may be forced to leave their homeland, rendered inhospitable by the climate or increasingly frequent catastrophic events, moving towards other territories of the globe [27], with all the consequences that large-scale migrations can produce. At the same time, lands that are currently inhospitable due to excessively cold temperatures could become habitable, unveiling new possibilities for human colonization. The poles and the regions closest to them could become the focus of renewed attention and new conquest projects.

Throughout the planet's history, the climate has changed many times, even due to sudden natural events. Never before, however, has human activity had such a significant influence in altering its balance. The debate on the role of human activity on climate and the environment has been ongoing for years, and, due to the lack of sufficient expertise, we do not feel equipped to participate in it. However, it is evident from reliable statistical data and scientific studies that the planet is undergoing a phase of profound climatic changes. The near future, for these reasons, could reveal scenarios that are difficult to imagine, even on the global political stage. But considering the vast scope of the phenomena and events discussed, particularly in terms of human security, what we can fear is that risks and dangers may increase, leading to a potential worsening for many people.

For these reasons, it is clear that in the coming years, climate will become an increasingly strategic factor for international stability and security, one that every global player will need to consider. In particular, certain areas of the planet—such as the Mediterranean, given its specific conditions—may face increasingly serious threats arising from climate change.

The technological revolution we are living through may certainly help us counter and mitigate the effects of climate change and address the dangers arising from environmental crises. However, it is also already a subject of competition among major global players. The need for greater cooperation among individuals and states to address the risks arising from climate change could, in fact, collide with the rivalries and tensions that these same phenomena might produce or exacerbate. Climate change and the technological revolution are, today, the two greatest challenges, from every perspective, for humankind. How we manage to confront them will determine much of our future.

References

1. Definizione di "Sicurezza", Enciclopedia Treccani. https://www.treccani.it/enciclopedia/sicurezza/
2. Harper K (2019) Il destino di Roma, Einaudi
3. Petraeus D, Roberts A (2024) L'arte della guerra contemporanea. Dalla caduta del Nazismo al conflitto in Ucraina, Utet
4. Williams P, McDonald M (2018) An introduction to security studies. In: VV.AA. Security studies: an introduction. Routledge
5. Drezner DW (2024) How everything became national security, foreign affairs, 12 Agosto 2024. https://www.foreignaffairs.com/united-states/how-everything-became-national-security-drezner
6. Beck U (2004) La società del rischio. Verso una seconda modernità, Carocci, Roma
7. Beck U, La società del rischio, op. cit pag. 25
8. Goodman S, Baudu P (2023) Climate change as a "threat multiplier": history, uses and future of the concept, center for climate and security. https://councilonstrategicrisks.org/wp-content/uploads/2023/01/38-CCThreatMultiplier.pdf
9. NASA Global Climate Change, What is Climate Change? https://science.nasa.gov/climate-change/what-is-climate-change/
10. Wadhams P (2017) Addio ai ghiacci, Bollati Boringhieri
11. McDonald M (2023) The climate change–security Nexus. In: VV.AA. Climate change and conflict in the Pacific. Routledge

12. Vivekananda J (2022) Why climate change matters for human security, UN University and UN Environment Programme. https://collections.unu.edu/eserv/UNU:8836/UNUUNEP_Vivekananda_RHER.pdf
13. Goodman S et al (2007) National security and the threat of climate change, the CNA Corporation. https://www.cna.org/reports/2007/national%20security%20and%20the%20threat%20of%20climate%20change%20%281%29.pdf
14. World Health Organization (2024) Greater Horn of Africa (Djibouti, Ethiopia, Kenya, Somalia, South Sudan, Sudan, Uganda): WHO Health Emergency Appeal 2024. https://www.who.int/publications/m/item/ghoa-who-health-emergen-cy-appeal-2024
15. Casola C (2022) Sahel. Conflitti, migrazioni e instabilità a sud del Sahara, Il Mulino
16. Strengthening Global Cooperation Vital in Addressing Climate-Induced Migration, UN International Organization for Migration, 2024. https://www.iom.int/news/strengthening-global-cooperation-vital-addressing-climate-induced-migration-iom
17. A complex nexus, UN International Organization for Migration. https://www.iom.int/climate-action
18. Collier P, Hoeffler A (1998) On economic causes of civil war. Oxford economic papers. Collier P, Hoeffler A (2004) Greed and grievance in civil war. Oxford economic papers
19. Gilli A et al (2024) Climate change and military power: hunting for submarines in the warming ocean. https://tnsr.org/2024/03/climate-change-and-military-power-hunting-for-submarines-in-the-warming-ocean/
20. NATO Climate Change and Security Impact Assessment, 2024.https://www.nato.int/nato_static_fl2014/assetspdf/2024/7/pdf/240709-Climate-Security-Impact.pdf
21. Andrea Gilli et al, op cit
22. Intergovernmental Panel on Climate Change, z, 2023. https://www.ipcc.ch/report/ar6/syr/downloads/report/IPCC_AR6SYR_LongerReport.pdf
23. UN Department of Economic and Social Affairs, What is the city but the people. https://www.un.org/en/desa/%E2%80%9Cwhat-city-people%E2%80%9D
24. Katrina and Oil Prices, Council on Foreign Relations, 2005. https://www.cfr.org/interview/katrina-and-oil-prices
25. Grech-Madin (2021) Water and warfare: the evolution and operation of the water taboo. Int Secur
26. Casini E, Deiana F, Manganelli E (2024) Artico e Mediterraneo: tra geopolitica e ambiente, un destino comune. In: AA.VV Il Mediterraneo Globale. Luiss University Press
27. Vince G (2023) Il secolo nomade. Bollati Boringhieri

Chapter 3
Safeguarding Potential Biosecurity Issues in Agribusiness

Lijun Shang

Abstract Biosecurity in the agribusiness sector is an often neglected and forgotten area of biological and chemical security. Complicated world environments and fast developing science and technologies especially those in agribusiness sector have put security education and training up to a higher level of importance and therefore building up capability in training in this sector is an urgent but challenge work. In this paper, we firstly present the potential biosecurity issues in agribusiness sector and then address the gaps in biosecurity education in this sector and explore how to build up capability through collaborative works. Through using biosecurity training as a conceptual model to be replicated for chemical and radiological education in agribusiness, we hope to potentially work towards a general framework, so that this capacity can be integrated into building up an International Biological Security Education Network (IBSEN) to tackle other CBRN issues in general.

3.1 Introduction

Economic development and social welfare in coming decades will be heavily dependent on advances in biotechnology, especially with the 4th industrial revolution. Following the devastating COVID-19 outbreak, State Parties increasingly want to be assured that biotechnology-driven development can take place safely and securely. Biosecurity issue has therefore been escalated to the highest level. State Parties and Biological Weapon Convention (BWC) have started to take measures to address this urgent need, such as constantly publishing and updating national biosecurity strategy, establishing working group, having focused discussion, etc.

Biosecurity not only limit to human but also apply to food, animal, plant and environment. This is so called agribusiness. However, biosecurity in the agribusiness sector is an often neglected and forgotten area. Vulnerabilities within the agribusiness

L. Shang (✉)
Biological Security Research Center, School of Human Sciences, London Metropolitan Universities, London, UK
e-mail: l.shang@londonmet.ac.uk

© The Author(s), under exclusive license to Springer Nature Switzerland AG 2025
M. F. Buongiorno et al. (eds.), *The Challenge of Environmental Security in the Euro-Mediterranean Region*, NATO Science for Peace and Security Series C: Environmental Security, https://doi.org/10.1007/978-3-031-99646-7_3

sector pose significant global risks to the population, and the ever-growing threat of climate change, the increasing presence of national and international conflicts and unstable geo-political environments in addition to advancements made in science and technology, draw great attention to the serious risks that have and are arising to the agribusiness sector [1]. Rapid advancements in science and technology hold significant advantages but also pose great risk as such advancements have allowed technological and biochemical manipulations to be easier than ever before. These advancements lead to questions about the presence of Dual-Use Research of Concern (DURC) and Dual-Use technology in agribusiness. Moreover, the specific biosecurity vulnerabilities which pose threats to the agribusiness sector can be exploited for harm by those with hostile intent, resulting in various consequences for Public Health (Physical and Psychological) and the Environment and thus disrupting states' society and economy.

3.2 Current Biosecurity Issue in Agribusiness

The term Agribusiness, as defined by David and Goldberg, is 'all operations involved in the cultivation, production, farm operations, processing and distribution of agricultural commodities and goods derived from them' [2]. Biosecurity, as defined under the Biological and Toxin Weapons Convention (BTWC), is the 'protection, control, and accountability measures implemented to prevent the loss, theft, misuse, diversion or intentional release of biological agents and toxins, and related resources as well as unauthorized access to, retention, or transfer of such materials' [3].

When discussing the specific biosecurity risks or vulnerabilities to the agribusiness sector it is useful to subcategorise these into Animals, Food, and Plant and Environment Biosecurity. It is also important to assess how these subcategories interlink and how attacks to one area of agribusiness may produce a 'domino effect' on the entire supply chain.

3.2.1 Food Security

Food biosecurity is dependent on the whole food supply chain, involving the entire agribusiness sector. Biosecurity in the agribusiness sector relates to the prevention, detection, and management of intentional and unintentional contaminations or attacks. An intentional attack in the agribusiness sector using Biological or Chemical Weapons would be classified as Agroterrorism which is a subset of Bioterrorism. Agroterrorism specifically targets livestock and crop production. If, however, the deliberate acts are motivated by financial or personal gain, they may be classed as an Agro crime. The specific act of protecting food from intentional threats is classed as Food Defence. Food Defence, as defined by the Global Food Safety Initiative is the 'process to ensure security of food and drink, food ingredients, feed, or food packaging from all forms of intentional malicious attack including ideologically motivated attacks leading to contamination or unsafe product' [4].

Food security may be threatened due to weather, lack of input of resources, diseases, national or international conflicts, and threats from intentional or unintentional actions. Food security is dependent on the whole food supply chain, incapsulating the entire agribusiness sector. As such, threats to one area of agribusiness may produce a domino effect on the entire food supply chain. Disruption to the food supply chain can disarray a state's socioeconomic stability, creating public distrust of the government and sparking questions of preparedness for managing biohazards, quality control and food safety. Food supply chain infiltrations encourage fear, expose states' vulnerabilities, and demonstrate considerable mental health effects on citizens.

Disruptions to food supply chains, and consequently the entire agribusiness sector may disproportionately affect low-income countries of which a large portion of their Gross domestic product (GDP) is from the agriculture industry. Other factors which make a country vulnerable is economic dependence on cash crops, insufficient self-sufficiency for food items, low farm biosecurity combined with intensive farming and few systems for monitoring the agribusiness sector.

3.2.2 Biosecurity in Animals, Plants and Environments

Animals found in the agriculture industry are highly susceptible to infectious diseases. This can be due to an absence of herd immunity caused by lack of prior exposure or vaccinations. Furthermore, animals which travel long distances may be immunocompromised due to heightened stress levels for extended periods of time. An absence of herd immunity in addition to immunocompromised animals increases their vulnerability to intentional or unintentional attack or contamination. The long-distance travel of animals means that a pathogen may easily disperse over a large area before animals display symptoms. Moreover, the long-distance movement further increases possibilities of pathogen dispersal from the original infection site.

Plants and crops are becoming increasingly vulnerable to intentional or unintentional attack or contamination due to large farm sizes and homogenous crop distribution, leaving large concentrations of plants or crops susceptible to pathogens. The use of crop varieties with narrow genetic bases further increases vulnerability to pathogens. The movement of crops across various regions has led to increased risk of exotic diseases. The increased need of agricultural produce to feed the world, makes it necessary to monitor potential disease outbreaks. Agricultural experts may have to grow experience and knowledge in the identification and management of the aforementioned diseases and thus increased biosecurity education is needed.

3.2.3 Effects of Climate Change

The twenty-first century is facing significant challenges regarding biosecurity of which are influenced via the evident consequences of climate change, advancements in science and technology, and the rising demands by the increasing populations of requiring the development of sustainable methods to increase food supply. Biosecurity vulnerabilities within the agribusiness sector may exhibit a range of societal (physical and psychological), environmental, economic and political consequences. For example, the United Kingdom and Europe are facing the re-emergence of various zoonotic and plant diseases (Bluetongue, Black stem rust etc.), in-addition to reaping the consequences of climate change, such as increased incidences of crop failures due to weather shocks. The challenge of sustainably increasing food supply to keep in-line with rising populations, has also led to various debates regarding Dual-use research of Concern such as genetic editing, etc.

Climate Change is having an increasing effect on crops and subsequently food production and poses a series threat to Global Food Security. Climate change can cause changes in air and ground temperature, air jet streams, wind, and may exhibit 'extreme weather'. Extreme weather is defined as "an event that is rare at a particular place and time of year". Extreme weather may present itself via sudden or dramatic changes in temperature, resulting in heat waves or cold snaps on land or sea, extreme rain resulting in severe floods, extreme droughts and wildfires, and extreme storms, such as hurricanes, typhoons and tornados.

Extreme weather as a direct consequence of climate change has been seen globally. Researchers from Climate Central, analysed the influence of human-caused climate change on heat waves over a 12-month period (May 2023–May 2024) using the Climate Shift Index. The study revealed that over 6.3 billion people experienced an extra 31 days of extreme heat compared to 1991–2020 and suggested that this rise in temperature was at least two times more likely due to human-caused climate change. Climate Central using the World Attribution criteria, identified 76 extreme heat waves over 90 different countries (Climate Change—Red Cross Red Crescent Climate Centre, 2024).

Weather shocks or extreme weather have disasters consequences to crop production. The rise in weather shocks has resulted in challenging decisions to be made by farmers, of which is more economically beneficial to harvest or abandon the fallen crop. Crop abandonment occurs where the value of the crop is equal or lower than the cost of harvesting. Crop failure may be defined as the total loss of crops and typically occurs due to weather shocks (spurred on via climate change) which leads to the destruction of crops by temperature changes, floods, droughts or an influx of pests. The relationship between crop failure and crop abandonment is a relatively new concept influenced by climate change. There are a multitude of factors which influence crop abandonment, and "Agricultural land abandonment is a complex, multidimensional and non-linear global phenomenon driven by multiple factors and with contrasting impacts on Agro-biodiversity, environment and society". It is also important to discuss that increases gaps between distributions of wealth and rising global conflicts will influence farmers decisions in crop abandonment.

Temperature changes may be catastrophic to crops, particularly during critical pheno-phases. Changes in temperature may affect crops development and growth and overall yield. Slight increases in temperature may be beneficial for crops located in Northen Regions, however, this is all dependent on the exact increase in temperature and strength of Ultraviolet light. Increases in temperature also result in increased water demand which leads to additional water stress.

Climate Change also affects the life cycle and behaviour of agricultural pests. Temperature changes, rises in CO_2, changes in rainfall all affect the behaviour patterns of pests. Temperature is a significant factor of which affects the behaviour and insect population dynamics, and rapid changes in temperature (extreme weather) may ‘trigger’ the expansion of pests/insects’ geographical range. In-addition, the introduction of initiatives to reduce the use of chemical/hazardous pesticides further exacerbates the needs for a sustainable method of pest control.

3.3 Biosecurity Education in Agribusiness

Due to the disastrous consequences which may arise when biosecurity is neglected within the industry or in the event of an unintentional or intentional shock event including climate change, natural catastrophic events and environmental crises, it is of great importance to educate the people involved, and also future experts in the field. Awareness raising and education of biosecurity within the Agribusiness sector is as equally important as any other biosecurity area. Civil society certainly can and should play an important role. Promotion of responsible biological and chemical science through training, education and awareness, including in relation to Dual Use Research of Concern is one of the effective measures to achieve Counter Proliferation Programme Strategical goals, which is the merit of NATO Science for Peace and Security Programme.

Complicated world environments and fast developing science and technologies including those in agribusiness sector have put security education and training up to a higher level of importance. Although State Parties work hard under Chemical, Biological, Radiological and Nuclear (CBRN) Conventions, adequate education and training for all stakeholders in the agriculture and food industry is not in place. Therefore, building up capability in training in this sector particularly in agribusiness sector is a challenge work.

In 2022, we started a new initiative to build up an International Biological Security Education Network (IBSEN), a similar educational mechanism to those developed for chemists under the Chemical Weapons Convention (CWC) and for nuclear scientists by the International Atomic Energy Agency (IAEA) and we hope this helps to provide a useful implementing method in biosecurity education under the BTWC.

However, there has been no systematic review of biosecurity education projects particular in agribusiness sector and what happened to them—apart from the fact that they have been sporadic, not highly funded, and importantly that few have been sustained over a long period or successfully replicated [5, 6]. Recently we have been awarded a priming pump grant from British Academy to focus on preparing a bid to Horizon Europe in 2024 to particularly address biosecurity education in the agribusiness sector, which is a less focused on, or often forgotten, area in CRBN security. We subsequently carried out a survey and organised two focused workshops. However, we found the current situation of biosecurity education in agribusiness sector is quite alarming.

In our recent survey [6], we found that there is a great gap in biosecurity education in the agribusiness sector across the European Universities. Biosecurity education with the Agribusiness sector at Bachelor and Master taught level is severely neglected. Only a tiny fraction of 4% of the total we surveyed (199 university courses) contained clear Biosecurity elements. Of this 4%, no courses addressed Agroterrorism, Agro-crime or Bioweapons, including the economical, physical, psychological and political implications. There is an urgent need for Biosecurity education at Universities within the Agribusiness Sector, and beyond.

We therefore argue for the development of an internationally coordinated curriculum which embeds biosecurity concepts into established courses. This will hugely enhance existing topics taught in courses to meet the urgent need in new circumstances derived by new science and technology.

In the meantime, the situation of lack of funding is very likely to change, and larger scale funding is going to become available to meet this critical gap in biological security. This is reflected in the Horizon Europe call, specifically the targeted Call and Topic in Pillar 2 is "Prevention, detection, response and mitigation of chemical, biological and radiological threats to agricultural production, feed and food processing, distribution and consumption".

Under our pump priming grant from the British Academy, after two successful workshops (one in London in June 2024 and another in Amsterdam in August 2024) we have built a team to propose a project to address the gaps in biosecurity education in Agribusiness sector. In this project, we aim to explore how to build up capability in training in this sector, which spans across the continents through collaborative works in several subpackages delivered by members of this consortium. We will analyse and compare existing security education materials and methodologies used in CBRN to further identify potential biosecurity material for the use in the Agribusiness sector, particularly related to the concerns caused by the fast advances in science and technology applying to agriculture and food industry, such as AI technology, gene editing, etc. We will then develop education materials, which will be done under the premises of "Prevention, detection, response and mitigation of chemical, biological and radiological threats to agricultural production, feed and food processing, distribution and consumption" as a critical infrastructure and risks pertaining therein. In developing and implementing these courses we will take advantage of our continuing work on biosecurity education, both in the recent production of a multiauthor textbook on biosecurity education (Essentials of Biosecurity: A Global Framework published by Wiley in Spring 2024) and our previous work on the efficient and effective use of

Team-Based Learning (TBL) exercises, designing of cartoons and podcasts, workshop, seminars, and simulation, etc. for teaching biosecurity education to different targeted stakeholders including university students, academics, professionals, industrials and policy makers at all levels. We hope through our proposed courses and relevant promotion workshops during their development and implementation, we will be able to highlight possible preparedness, mitigation and response plans for national authorities and the private sector.

We are therefore using biosecurity training as a conceptual model to be replicated for chemical and radiological education in agribusiness. Investing in education is expected to bring returns in awareness, security, risk mitigation (Prevention, detection, response and mitigation of threats).

3.4 Summary

Biological security requires prevention of natural, accidental, and deliberate disease. A fundamental problem for future biosecurity is that the scientists creating the revolution in the life sciences have little understanding of the problem of dual-use and biosecurity. This is a big education gap.

There are several initiatives from civil society with aim to closing this major gap in biosecurity, such as our recent efforts in "developing an International Biological Security Education Network (IBSEN)" to support the Tianjin Guidelines for Codes of Conduct under the BTWC and WHO framework, and more recently "scoping biological security risks and biosecurity education for a resilient agribusiness sector" to specifically address the potential biosecurity issues in agribusiness. Through these projects, we hope to potentially work towards a general framework, so that this capacity can be integrated into the applicant's recent initiative of building up an international biological security education network to tackle other CBRN issues in general. From across the spectrum to reduce the chances and impact of another COVID-19, foot and mouth or Dutch elm diseases, we hope to bring together experts and policymakers who have worked, separately on flu, COVID-19, monkey pox, foot and mouth and other diseases that affect humans and animals. This will help us learn from past epidemics and plan for the future while reducing inequalities from the outset.

These projects also aim to develop a platform consisting of biosecurity experts with delegates from all Member States which may foster a web of policies, approaches and measures that strengthen biosecurity within Member States. However, sustainable support from all stakeholders is extremely needed to safeguard this fast developing and improving area.

Acknowledgements The author thanks for Miss Olivia Ibbotson for her initial works on biosecurity in agribusiness, and funds from Joseph Rowntree Charitable Trust, British Academy, and London Metropolitan University.

References

1. Ibbotson, Shang (2024) An overview of biosecurity risks posing threat to the agribusiness sector in the 21st century. Front Gene Editing
2. Rust et al (1957) A concept of agribusiness. J Farm Econ 39:1042. https://doi.org/10.2307/1234228
3. EU Working Paper (2022) Ninth review conference of the biological and toxin weapons convention—1123_EU Working Paper on Biosafety and Biosecurity Final to be submitted to BTWC RevCon.pdf (unoda.org)
4. Spink JW (2023) The GFSI food fraud prevention compliance development and requirements: a ten-year review. Trends Food Sci Technol 138:766–773
5. Shang et al (2022) Key issues in the implementation of the Tianjin biosecurity Guidelines for codes of conduct for scientists: a survey of biosecurity education projects. Biosafety Health. https://doi.org/10.1016/j.bsheal.2022.08.003
6. Ibbotson, Shang (2024) The urgent need for biosecurity education at Universities within the agribusiness sector. J Biosafety Biosecur

Chapter 4
From Ocean Currents to Power Struggles: Climate Change and Emerging Geopolitical Threats in the Arctic

Sandro Carniel, Andrea Gilli, Giuliana Panieri, Marco Dordoni, and Mauro Sclavo

Abstract This paper explores how climate change transforms the Arctic from a remote, ice-locked frontier into a critical arena where environmental fragility meets geopolitical volatility. As ocean currents shift and sea ice retreats, the Arctic emerges as both a bellwether of planetary change and a stage for intensifying strategic rivalries. The ocean and climate are deeply interconnected systems, and their evolving dynamics are reshaping global power relations—nowhere more visibly than in this rapidly warming region. New maritime routes and resource extraction opportunities are redrawing strategic maps, as Arctic states and global powers reposition themselves in response to shifting environmental and security landscapes. These changes, although cantered in the Arctic, reverberate through interconnected climate systems, impacting weather patterns and maritime security as far as the Euro-Mediterranean region. The paper analyses the complex interplay between ecological transformation and great power competition through an interdisciplinary lens that combines climate science, international relations, and security studies, this paper analyzes the complex interplay between ecological transformation and great power competition. In doing so, it calls for integrated, multilateral governance solutions capable of reconciling ecological protection with economic ambition and emerging security imperatives.

Keywords Arctic climate transformation · Arctic geopolitcs · Climate-Security Nexus

S. Carniel (✉) · G. Panieri · M. Sclavo
Institute of Polar Sciences, National Research Council, Venice, Italy
e-mail: sandro.carniel@cnr.it

A. Gilli
School of International Relations, University of St. Andrews, St. Andrews, Scotland, UK

M. Dordoni
University for Foreigners, Perugia, Italy

© The Author(s), under exclusive license to Springer Nature Switzerland AG 2025
M. F. Buongiorno et al. (eds.), *The Challenge of Environmental Security in the Euro-Mediterranean Region*, NATO Science for Peace and Security Series C: Environmental Security, https://doi.org/10.1007/978-3-031-99646-7_4

4.1 Introduction

The interdependence between the ocean and the climate system—often referred to as the ocean-climate nexus—is central to understanding the profound environmental and geopolitical transformations of our time [1]. Oceans regulate Earth's climate by absorbing heat and carbon dioxide, driving weather patterns through complex current systems, and influencing sea level and ice cover.

At the same time, climate change is disrupting these processes, altering ocean circulation dynamics, accelerating sea ice melt, and triggering feedback loops that affect global climate stability.

Nowhere is this nexus more evident—and more consequential—than in the Arctic.

The Arctic is warming nearly four times faster than the global average, making it one of the most rapidly changing regions on Earth [2]. Sea ice is retreating at unprecedented rates, permafrost is thawing, and new oceanographic regimes are emerging [3].

These physical changes are not occurring in isolation: they are reshaping the geopolitical landscape. As previously ice-bound waters become seasonally navigable and new resources become accessible, the Arctic is transforming from a frozen frontier into a contested strategic space. Coastal states, global powers, and non-Arctic actors are increasingly asserting interests in the region, motivated by access to untapped hydrocarbons, fishing grounds, critical minerals, and shorter shipping routes.

This convergence of environmental fragility and geopolitical interest elevates the Arctic to a critical site of inquiry. Climate change is not only a driver of physical transformation but also a catalyst for political and economic reconfiguration. The emerging landscape presents both risks—such as heightened military presence and contested claims—and opportunities for multilateral cooperation on governance, environmental protection, and sustainable development.

It is worth recalling that changes occurring in the Arctic ocean circulation patterns and strategic maritime routes are increasingly linked to security concerns, energy dependencies, and geopolitical stability in the Euro-Mediterranean region.

This paper aims to trace the link between physical changes in the Arctic Ocean and the evolving geopolitical dynamics they engender.

Through an interdisciplinary approach that draws from oceanography, climate science, and international relations, it will examine how the transformation of ocean currents, sea ice extent, and marine ecosystems is giving rise to new strategic considerations. In doing so, the paper seeks to contribute to a broader understanding of how climate-induced environmental change is reshaping not only the physical world, but also the architecture of global power.

4.2 Understanding the Ocean-Climate Nexus

The ocean and the climate system are intricately linked through a series of physical, chemical, and biological processes that regulate the Earth's energy balance and carbon cycle. Covering over 70% of the planet's surface, the oceans act as a vast buffer against atmospheric change, playing a critical role in modulating temperature, storing carbon dioxide, and redistributing heat across the globe. Understanding this ocean-climate nexus is essential to grasp the broader implications of climate change, particularly in sensitive regions like the Arctic.

One of the ocean's most fundamental climate-regulating functions is their capacity to absorb and store heat. Since the beginning of the industrial era, the ocean has absorbed over 90% of the excess heat trapped by greenhouse gases in the atmosphere [4]. While this heat uptake has delayed the full manifestation of global warming on land, it has also triggered thermal expansion, sea level rise, and changes in marine ecosystems. The upper layers of the ocean, particularly the top 700 m, have warmed significantly in recent decades, altering the vertical mixing and stratification that influence nutrient availability, biological productivity, and carbon exchange [5].

In addition to heat, the ocean acts as a major carbon sink, absorbing around 25–30% of anthropogenic CO_2 emissions annually. Through both physical and biological processes, carbon dioxide is transported from the atmosphere into the ocean interior, where it can be stored for centuries, either dissolved in cold waters which then sink to the deep ocean, or via the photosynthetic activity of phytoplankton and the subsequent sinking of organic material. However, as ocean temperatures rise and acidification intensifies, the efficiency of these processes may decline, undermining one of the planet's most crucial climate stabilizing mechanisms.

Ocean currents are another key component of the climate system. Large-scale circulation patterns—such as the thermohaline circulation (also called the global conveyor belt)—transport warm water from the tropics toward the poles and return cold, dense water equatorward at depth. This circulation not only redistributes heat but also influences weather systems, precipitation patterns, and sea ice formation. Central to this process is dense water formation, particularly in polar regions, where cold, salty water sinks and drives deep ocean currents. In the Arctic, the formation of dense waters in the Nordic Seas and in the Labrador Sea is vital for the Atlantic Meridional Overturning Circulation (AMOC), a critical ocean current system that helps regulate the global climate by transporting warm water toward the poles and cold water toward the equator.

However, the ocean-climate system is governed by nonlinear feedback mechanisms that can amplify or accelerate change. One of the most significant in the Arctic is the albedo effect: as sea ice melts, it exposes darker ocean surfaces that absorb more solar radiation, leading to further warming and more ice melt—a self-reinforcing loop. Similarly, the influx of freshwater from melting glaciers and permafrost disrupts salinity balances, potentially weakening the thermohaline circulation and altering regional and global climate patterns [6, 7].

These feedbacks introduce uncertainties and tipping points into the climate system. The Arctic, with its shallow marginal seas and strong seasonal contrasts, is particularly sensitive to such shifts.

Such disruptions above described, including the AMOC one, could impact climate stability and precipitation regimes in Europe and the Mediterranean basin, influencing agricultural productivity, water security and energy supply stability [8, 9].

Understanding how oceanic processes are responding to and reinforcing atmospheric changes is therefore central not only to predicting future climate trajectories, but also to anticipating the geopolitical consequences of a rapidly transforming marine environment.

In sum, the ocean-climate nexus is not a passive backdrop to global change, but an active and dynamic driver. Its complexities must be accounted for in both scientific modeling and in policy frameworks that aim to manage climate risks and their cascading effects across environmental and geopolitical domains.

4.3 The Arctic as a Sentinel of Change

The Arctic is often described as the Earth's early warning system—a sentinel of change—because the effects of climate change manifest here earlier, faster, and more dramatically than anywhere else on the planet. Warming in the region is occurring at nearly four times the global average [2], a phenomenon known as Arctic amplification, driven by complex interactions between the atmosphere, ocean, ice, and land. These changes are not merely local anomalies, they have far-reaching consequences for the global climate system, ocean circulation, ecosystems, and human societies [10].

Among the most visible and consequential transformations is the rapid loss of sea ice. Summer sea ice extent has declined by approximately 13% per decade since the late 1970s [6], and multi-year ice—older, thicker, and more resilient—has largely been replaced by thinner, first-year ice that melts more readily [11]. This reduction not only alters surface reflectivity (albedo), accelerating regional warming, but also destabilizes weather patterns beyond the Arctic, potentially influencing the jet stream and mid-latitude weather extremes [12]).

In addition, Arctic surface temperatures are rising at an unprecedented pace [6]. This warming contributes to the thawing of permafrost [13], the frozen layer of soil and organic material that underlies much of the Arctic landmass. As permafrost thaws, it releases greenhouse gases—particularly methane and carbon dioxide—that have been trapped for millennia, creating powerful feedback loops that further exacerbate global warming. The degradation of permafrost also threatens infrastructure, ecosystems, and the traditional ways of life of Indigenous communities [14].

Resulting changes in ocean circulation patterns are emerging as another critical consequence. The inflow of freshwater from melting ice and increased river discharge is altering salinity levels and stratification in Arctic waters, with potential downstream effects on deepwater formation and the AMOC. If disrupted [15], these currents could profoundly affect climate stability in North America and Europe, including the Euro-Mediterranean region. Additionally, the Arctic Ocean itself is undergoing ecological shifts, including altered nutrient dynamics, changing species distributions, and the northward migration of sub-Arctic marine life [16].

These environmental transformations carry deep consequences for local ecosystems and coastal communities. Arctic biodiversity—already adapted to extreme conditions—is under stress from habitat loss, warming waters, and shifting food webs [17]. Species such as polar bears, walruses, and Arctic cod face growing survival challenges. Meanwhile, Indigenous peoples whose livelihoods depend on predictable sea ice and seasonal rhythms face cultural, economic, and existential threats. Traditional navigation routes become hazardous, and food security is jeopardized as subsistence hunting and fishing patterns are disrupted.

Paradoxically, the same processes that endanger the Arctic environment are also opening new economic and geopolitical opportunities [18]. The decline in summer sea ice is making routes that were once difficult to navigate increasingly accessible, e.g. the Northern Sea Route along Russia's Siberian coast. More importantly, once-impassable routes such as the Northwest passage through Canada's Arctic Archipelago and the Transpolar routes, are now becoming navigable for an increasing number of weeks each year. These shortcuts can reduce transit times between Europe and Asia by thousands of kilometers, making them attractive alternatives to traditional routes via the Suez or Panama Canals.

However, the prospect of increased shipping traffic, resource extraction, and tourism traffic poses serious environmental risks [19] to the Arctic ecosystems, particularly vulnerable to oil spills, underwater noise, and invasive species introduced by ballast water. Limited infrastructure and extreme conditions hinder emergency response capabilities, meaning that even minor accidents can have outsized ecological consequences. Moreover, black carbon (soot) emissions from vessels and heavy machinery darken ice surfaces, accelerating melt and contributing to further warming [20].

In addition, the emergence of new Arctic shipping lines could potentially alter global maritime trade patterns, impacting established Mediterranean routes, notably through the Suez Canal, thereby reshaping regional economic and strategic balances.

In short, the Arctic is both an indicator of planetary health and a precursor of emerging risks and opportunities. Understanding its transformations is essential not only for protecting a fragile environment, but also for anticipating the broader geopolitical shifts that these changes are likely to trigger.

4.4 Great Power Competition in the Arctic

In addition, and perhaps more importantly, the Arctic is increasingly becoming a stage for intense geopolitical competition among the world's great powers. As climate change accelerates the melting of sea ice, it is opening up new navigable routes, access to untapped resources, and potential strategic military advantages.

This section explores the conflicting national interests, strategic motivations, and economic implications that are shaping the Arctic's future.

4.4.1 *Conflicting National Interests*

The primary players in the Arctic are Russia, the United States, Canada, Norway, and Denmark/Greenland, each with distinct national interests and competing territorial claims. For Russia, the Arctic is of immense strategic importance, with vast oil and gas reserves, as well as access to newly accessible sea routes. Russia has been increasingly militarizing its Arctic regions, building new military bases [21], and strengthening its presence in the region. The United States, with its longstanding interests in maintaining freedom of navigation and security in the region, views Russia's military expansion in the Arctic as a challenge to its dominance in the North [22]. Canada, with its extensive Arctic coastline, is focused on protecting its sovereignty and maintaining its claim over the Northwest Passage, which Russia and others view as international waters [23]. Norway is a key NATO member and is concerned with its own territorial claims and security in the Barents Sea, while Denmark (via Greenland) seeks to assert its claim over the extended continental shelf and its potential resources. These conflicting national interests are exacerbated by the need to control the region's emerging shipping routes and the exploitation of natural resources, making the Arctic a strategic focal point for all parties involved [24].

4.4.2 *Strategic Relevance of Arctic Region*

The strategic significance of the Arctic is huge. For Russia, the Arctic represents a vital corridor for military defense, energy resources, and economic development. Russia's ambitions in the Arctic are driven by its desire to secure a permanent military and economic presence in the region, particularly as melting ice makes it more accessible [25].

In this context the Arctic region assumes a pivotal role in restoring Russia's status among the world's great powers, both economically and geopolitically. The expansive Russian Arctic is exceptionally rich in natural resources, mainly oil and gas, and has thus become the focus of ambitious initiatives such as the Yamal LNG 1 and 2 projects. [26, 27]

These schemes aim not only to exploit hydrocarbon reserves but also to develop port and airport infrastructures along the northern coast, thereby enhancing the navigability of the Northern Sea Route. As the region's ice cover recedes at an unprecedented rate, this shipping corridor can halve transit times and fuel consumption on voyages between Asia and major Central European ports like Rotterdam and Hamburg. Originally launched in the 2010s with significant European participation—particularly from France and Germany—the projects saw their Western financing freeze following the outbreak of hostilities in Ukraine; they are now principally underwritten by Chinese and Indian capital [28]. To safeguard these substantial economic interests and assert control over the Arctic passage, the Russian Federation has reactivated and strengthened its military presence, reviving a defensive network that lay largely dormant after the Cold War.

Similarly, the U.S. sees its Arctic interests through the lens of national security, aiming to ensure the region remains free from adversarial military buildup while maintaining freedom of navigation. NATO, of which many Arctic nations are members, is deeply concerned with the balance of power in the region and the threat posed by Russia's growing presence. NATO countries, particularly those with Arctic territory such as Norway, are strengthening their military infrastructure in response to Russia's increased activity [29]. The strategic importance of controlling Arctic waters for naval power projection, as well as access to the region's natural resources, has led to an intensification of military posturing in the region [30].

The Arctic is also becoming a focal point for non-Arctic nations, most notably China. China has expressed a growing interest in the region, which it sees as a key part of its broader geopolitical strategy. China's "Polar Silk Road" initiative aims to establish a trade route linking China to Europe through the Arctic, capitalizing on the shorter shipping times and increased access to resources. As an observer in the Arctic Council, China is working to increase its influence in the region, despite not having any direct territorial claims [31].

The European Union (EU), while not a direct player in terms of territorial claims, is deeply concerned with the environmental impact of Arctic exploitation. The EU has championed climate change initiatives and seeks to ensure that Arctic development is conducted sustainably [32]. However, its limited leverage in the region, given the competing interests of Arctic states, means that its influence over Arctic governance and development is constrained. In addition, while promoting a robust environmental agenda in the Arctic, the EU also recognizes the strategic necessity of safeguarding resource access and maritime security, factors directly tied to stability in the Euro-Mediterranean region.

4.4.3 Economic Implications (Trade, Shipping, Access to Untapped Resources)

One of the main drivers of competition in the Arctic is the region's wealth of untapped natural resources. The Arctic seabed is estimated to contain significant reserves of oil, gas, rare earth elements, and minerals [33]. According to some estimates, the Arctic could hold up to 30% of the world's undiscovered gas reserves and 13% of its untapped oil resources [34]. These reserves are particularly attractive to countries like Russia and the United States, which are keen to secure energy independence and maintain economic growth.

European nations such as Italy, France, Germany heavily reliant on secure energy imports through Mediterranean corridors, increasingly view Arctic resource competition as directly relevant to their long-term security strategies [35].

Furthermore, with the Arctic's receding ice cover, previously inaccessible areas are becoming more viable for exploration and exploitation. As a result, mapping the Arctic seabed and securing exclusive rights to exploit these resources has become a key aspect of the geopolitical rivalry in the region.

The melting ice is opening up new shipping routes, notably the Northwest Passage and a series of Transpolar routes. These routes, which were previously closed or extremely difficult to navigate, are now offering shorter, more efficient passages for global trade. This is endangering the pivotal role of countries like Russia, which stands to largely control over the Northern Sea Route, once the only feasible navigable option. By reducing the distance between Europe and Asia, these new routes are revolutionizing global trade, offering faster shipping times and cost savings.

However, as more shipping companies consider these routes, concerns arise over the environmental impact. The increased traffic could lead to oil spills, maritime accidents, and the introduction of invasive species, posing risks to both the environment and regional security.

In addition to shipping, the region's untapped resources—oil, gas, rare earth minerals, and fisheries—present further economic opportunities. As these resources become increasingly accessible, nations are looking to assert control over them through both bilateral negotiations and legal frameworks. The global demand for rare earth elements, for instance, is driving competition in the Arctic, as these minerals are vital for electronics, renewable energy technologies, and defense systems.

4.5 Geopolitical and Security Threats

The Arctic region is undergoing a profound transformation driven by climate change, which is not only altering the physical landscape but also reshaping geopolitical dynamics. As sea ice retreats, new maritime routes are emerging and access to previously untapped natural resources—hydrocarbons, minerals, and fisheries—is becoming increasingly feasible. These developments are attracting the attention of both Arctic and non-Arctic states, prompting a significant shift in the region's strategic calculus. Once characterized as a zone of low tension and cooperative scientific engagement, the Arctic is now increasingly framed as a space of geopolitical contestation and multifaceted security threats. Addressing these complex and interrelated risks requires renewed diplomatic engagement, transparency in military postures, confidence-building measures, and the strengthening of multilateral governance frameworks such as the Arctic Council and UNCLOS. Without proactive efforts to manage competition and promote cooperation, the Arctic risks becoming not a zone of peace, but a flashpoint in an increasingly fragmented global order.

4.5.1 Military Presence and Strategic Competition

A key dimension of this shift is the marked increase in military presence and exercises by Arctic states. Russia, which controls approximately half of the Arctic coastline, has significantly expanded its military infrastructure since the early 2010s. This includes the reopening of Cold War-era bases, the deployment of S-400 missile systems, the enhancement of the Northern Fleet, and the construction of new radar installations and airstrips along the Northern Sea Route (NSR) [36]. The scale and tempo of Russian military exercises in the High North have also increased, projecting both deterrence and sovereignty claims.

From a strategic-security perspective, the Arctic represents both the most direct ballistic-missile corridor between the United States and the Russian Federation and a vital maritime passage to the Atlantic via the Greenland–Iceland–United Kingdom (GIUK) gap. For Moscow, this dual utility underpins its ability to threaten U.S. and NATO stability through increasingly sophisticated long-range strike systems—most notably the Kalibr missile family—capable of striking critical infrastructures on the North American mainland with high precision. The suspension of bilateral arms-control negotiations in the wake of the Ukraine conflict has further resurrected the specter of a renewed Russian nuclear threat, placing the Arctic squarely at the center of contemporary U.S. strategic discourse.

Indeed, recent political developments have elevated the High North to a linchpin of American foreign-policy priorities, with proposals for expanded radar networks and defensive installations across Canada and Greenland aimed at bolstering deterrence against emergent Russian capabilities. [25].

Within the NATO-Russia security calculus, attention has converged on Europe's High North, where the proximity of the Kola Peninsula—a few kilometers from Norway and Finland—magnifies the Allied vulnerability. Home to Russia's Northern Fleet and its second strategic nuclear base, the Kola Peninsula undergirds Moscow's so-called Bastion strategy, which leverages increasingly silent nuclear-powered submarines to penetrate the "Bear Line" and threaten transatlantic sea lines of communication via the GIUK gap.

In counter-response, NATO member states—particularly Norway, Denmark, Sweden, and Finland, alongside non-Arctic powers such as the United Kingdom, the Netherlands, and Germany—have intensified their Arctic defense posture and prioritized enhanced situational awareness, readiness, and force posture in the High North. The 2022 NATO Strategic Concept formally incorporated the region into Alliance planning, and the 2023 Vilnius Summit established dedicated "Regional Plans" under the operational authority of Joint Force Command Norfolk [37] (NATO Strategic Concept 2022, Vilnius Summit Communicate [38]).

The accession of Sweden and Finland as full NATO members, combined with the development of defensive infrastructures in Rovaniemi (Finland) and Kiruna (Sweden), exemplifies this shift. Operationally, the scale and complexity of High-North exercises have surged: the 2024 Nordic Response marked the largest Allied drill in the region since the Cold War, and preparations are already underway for a

2026 iteration expected to exceed its predecessor, underscoring the Arctic's growing centrality in NATO's strategic posture Boulègue M [39].

NATO, and particularly its Nordic members, is intensifying operational efforts to improve multi-domain readiness and defense capabilities by integrating advanced technologies, including AI and autonomous unmanned systems. Drawing from operational insights in Ukraine—specifically the tactical use and countermeasures of drone technologies—these adaptations are being tailored to the distinct environmental and strategic conditions of the High North theater.

4.5.2 Escalation Risks and Legal Disputes

The expansion of military assets in proximity to poorly defined borders increases the risk of conflict escalation, whether due to misperception, technical error, or political provocation.

Increasing militarization and hybrid threats in the Arctic have potential spill-over effects on Mediterranean NATO allies, compelling integrated strategic responses and reinforcing trans-regional defense cooperation.

Disagreements over maritime boundaries and interpretations of international law exacerbate these risks. For instance, while Canada and Russia assert sovereignty over the Northwest Passage and Northern Sea Route respectively, the United States and several allies consider these to be international straits subject to freedom of navigation under the United Nations Convention on the Law of the Sea (UNCLOS) [40]. These divergent legal positions may result in diplomatic tensions or even incidents at sea, particularly as commercial and military traffic intensifies.

4.5.3 Hybrid Threats and Infrastructure Vulnerability

In parallel with conventional military developments, the Arctic is increasingly exposed to hybrid threats, including cyber operations, disinformation campaigns, and the potential sabotage of critical infrastructure. Undersea communication cables, energy pipelines, and satellite systems in the region are vital for global connectivity and logistics, yet their remote locations and harsh conditions render them vulnerable [41]. Hybrid threats originating from Arctic tensions, including cyberattacks and disinformation, could exploit vulnerabilities in critical infrastructures across the Mediterranean, highlighting the interconnectedness of regional security.

Several analysts have warned that such hybrid threats are difficult to attribute and counter, making them a preferred tool for strategic competition below the threshold of armed conflict.

The information domain is another emerging battleground. Narratives around territorial rights, environmental stewardship, and Indigenous sovereignty are increasingly manipulated through strategic communication efforts, including state-sponsored media and covert influence operations. This information warfare complicates cooperative governance and can erode trust among Arctic stakeholders.

In light of the recent acts of sabotage in the Baltic Sea [42, 43]—most notably the disruption of undersea infrastructure between Estonia and Sweden—and the planned deployment of new submarine cable networks in the Barents Sea, NATO has identified undersea communications as a critical vulnerability [44] requiring enhanced collective preparedness. Allied efforts now encompass the development of comprehensive surveillance regimes (including increased maritime patrols and unmanned underwater vehicles), the establishment of rapid-response protocols for cable-repair operations, and the integration of cyber-defense measures to detect and mitigate hostile interference in real time. Moreover, NATO's forthcoming "Subsea Security Framework" seeks to codify best practices for member-state coordination, legal recourse, and engagement with civilian telecommunications providers to ensure resilience against both state-sponsored and non-state adversaries seeking to exploit these vital data conduits. Equally pressing is the protection of offshore energy installations along Norway's Arctic coastline, where oil and gas platforms constitute high-value nodes in Europe's energy security architecture. These fixed assets are inherently exposed to asymmetric threats—ranging from long-range precision strikes and underwater mining to cyber intrusions targeting platform-control systems—and their isolation in the High North complicates both deterrence and defense. In response, NATO and partner navies have intensified joint exercises emphasizing multi-domain anti-access/area-denial (A2/AD) countermeasures, enhanced coordination with the Norwegian Coast Guard, and the deployment of hardened command-and-control networks to maintain operational continuity even under sustained hostile action. Such initiatives reflect a broader strategic imperative: to safeguard the integrity of energy supply chains and maintain freedom of navigation in the increasingly contested maritime and subsea environments of the Arctic.

4.5.4 Environmental Insecurity and Human Dimensions

Finally, the Arctic faces growing environmental security challenges with profound societal implications. Rapid warming, permafrost thaw, and sea-level rise are accelerating the degradation of ecosystems and infrastructure, threatening the livelihoods of Indigenous and local communities. As a result, the region may witness forced migration and the displacement of populations, particularly in areas where governance capacity is weak or adaptation resources are lacking [45]. Moreover, resource scarcity, such as reduced fish stocks or freshwater availability, can exacerbate tensions within and between states, acting as a conflict multiplier.

These human security concerns are often overlooked in traditional security paradigms but are essential to a comprehensive understanding of Arctic geopolitics.

Integrating environmental, social, and economic dimensions into security assessments is therefore critical to formulating sustainable and peaceful policies in the region.

4.6 Conclusions

The Arctic is no longer a faraway, frozen frontier. It has become a dynamic region where climate change and geopolitical competition are coming together in powerful ways. As this paper has shown, the loss of sea ice, changes in ocean currents, and shifts in marine ecosystems are not just environmental problems—they are closely linked to changes in global power and strategy. The Arctic clearly shows how environmental fragility and political tensions can reinforce each other, creating a region where strategic interests and ecological risks are deeply connected.

This analysis highlights three main points. First, the physical changes in the Arctic Ocean are reshaping global strategies, opening new shipping routes, resource areas, and new spaces for international competition. Second, the connection between the ocean and the climate must be seen not only as a scientific issue but also as a driver of political and economic change. Third, if action is not taken early, the Arctic risks becoming a place of conflict instead of cooperation, with both the environment and security suffering as a result.

Considering these findings, there is an urgent need for coordinated action at the point where climate and security issues meet. Future governance must protect the environment, support sustainable development, and maintain strategic stability at the same time. Achieving this will require new forms of multilateral cooperation, linking climate science, international law, security policy, and indigenous knowledge into a unified and resilient system.

Future research must continue to build strong connections across disciplines, using the latest developments in oceanography, climate modeling, and security studies to better predict and manage new risks. Policymakers should focus on flexible, forward-looking strategies that can handle both environmental and political uncertainties. In the rapidly changing Arctic, peace and sustainability must be treated not as opposing goals, but as two sides of the same effort.

The Arctic is a warning for the rest of the planet. How the world responds to the profound changes happening there will be a critical test of our ability to manage the combined challenges of climate disruption and geopolitical competition.

References

1. Carniel S, Russo A, Lissi F (2024) Climate change and security workshop proceedings, conference proceedings, CMRE-CP-2024-001

2. Rantanen M et al (2022) The Arctic has warmed nearly four times faster than the globe since 1979. Commun Earth Environ 3(1):168
3. IPCC (2023) Climate change 2023: synthesis report. In: Contribution of Working Groups I, II and III to the sixth assessment report of the intergovernmental panel on climate change. https://www.ipcc.ch/report/ar6/syr/
4. Cheng L et al (2023) Another record: ocean warming continues through 2022 despite La Niña conditions. Adv Atmos Sci 40(2):249–260. https://doi.org/10.1007/s00376-023-2385-2
5. Bindoff NL et al (2019) Changing ocean, marine ecosystems, and dependent communities. In: IPCC special report on the ocean and cryosphere in a changing climate. https://www.ipcc.ch/srocc/
6. Intergovernmental Panel on Climate Change (IPCC) (2021) Climate Change 2021: the physical science basis. In: Contribution of working Group I to the sixth assessment report of the IPCC. Cambridge University Press. https://www.ipcc.ch/report/ar6/wg1/
7. Rahmstorf S, Box JE, Feulner G, Mann ME, Robinson A, Rutherford S, Schaffernicht EJ (2015) Exceptional twentieth-century slowdown in Atlantic Ocean overturning circulation. Nat Clim Chang 5(5):475–480. https://doi.org/10.1038/nclimate2554
8. Ciscar JC, Feyen L, Soria A, Lavalle C, Raes F, Perry M, Baranzelli C (2018) Climate impacts in Europe: final report of the JRC PESETA III project. Environ Impact Assess Rev 72:1–12. https://doi.org/10.1016/j.eiar.2018.05.007
9. Intergovernmental Panel on Climate Change (IPCC) (2022) Climate change 2022: impacts, adaptation, and vulnerability. In: Contribution of Working Group II to the sixth assessment report of the IPCC. Cambridge University Press. https://www.ipcc.ch/report/ar6/wg2/
10. Meredith M et al (2019) Polar regions. In: IPCC special report on the ocean and cryosphere in a changing climate
11. Stroeve J, Notz D (2018) Changing state of Arctic sea ice across all seasons. Environ Res Lett 13(10):103001
12. Cohen J et al (2020) Divergent consensuses on Arctic amplification influence on midlatitude severe winter weather. Nat Clim Chang 10:20–29
13. Turetsky MR, Abbott BW, Jones MC, Anthony KMW, Olefeldt D, Schuur EAG, McGuire AD (2020) Carbon release through abrupt permafrost thaw. Nature Geosci 13(2):138–143. https://doi.org/10.1038/s41558-020-0889-z
14. AMAP (2021) Arctic climate change update 2021. Arctic Monitoring and Assessment Programme
15. Li C et al (2021) Subpolar North Atlantic overturning and hydrographic changes during the last interglacial. Nature 593:583–587
16. Fossheim M et al (2015) Recent warming leads to a rapid borealization of fish communities in the Arctic. Nat Clim Chang 5:673–677
17. CAFF (2017) State of the Arctic Marine biodiversity report. Conservation of Arctic Flora and Fauna
18. Lackenbauer PW, Lalonde S (eds) (2017) Emerging legal orders in the arctic. Routledge
19. PAME (2022) Arctic Shipping Status Report 2022. Protection of the Arctic marine environment working group
20. AMAP (2015) Black carbon and ozone as Arctic climate forcers. Arctic Monitoring and Assessment Programme
21. Tisch A (2024) Russia's Military strategy in the Arctic: expanding influence and presence. Russ Def Rev 43(2):100–114
22. Borgerson SG (2019) The geopolitical implications of the Arctic's emerging shipping routes. Foreign Aff 98(2):70–79
23. Young OR (2020) The Arctic council and the challenge of sovereignty claims: the case of the northwest passage. Arctic Policy Rev 12(3):45–59
24. Hassel M (2021) The Arctic's strategic importance and international dynamics: resource access and territorial claims. J Arctic Secur 15(4):22–37
25. Jacobsen A, Barker M, Lewis P (2024) Melting ice and military buildup: the evolving strategic importance of the Arctic. Int Affairs Rev 37(4):75–92

26. Closson S (2017) Russia's Arctic ambitions and the Yamal LNG projects. Geopolitical Rev 25(1):45–62
27. Devyatkin A (2024) Economic development in the Russian Arctic: Hydrocarbon resources and infrastructure projects. Arctic Econ J 15(3):112–130
28. Moss S (2022) The shifting financial landscape in the Russian Arctic: From Western investment to Eastern capital. J Glob Finance 38(5):200–215
29. Anderson K (2023) Strategic military posturing in the Arctic: NATO's response to Russian activity. J Int Secur 42(2):153–168
30. Peterson J, Williams S, Smith A (2023) Naval power and resource access: the Arctic's growing military relevance. Naval Strat Q 39(1):51–67
31. Wu H (2023) China's polar Silk Road initiative and its influence on Arctic geopolitics. Asian Geopolitics Rev 9(2):129–145
32. EU Arctic Policy (2021) The European Union's approach to sustainable development in the Arctic. European Commission. Retrieved from https://europa.eu/arctic-policy
33. U.S. Geological Survey (2008) Circum-Arctic resource appraisal: estimates of undiscovered oil and gas north of the Arctic circle. https://pubs.usgs.gov/fs/2008/3049/fs2008-3049.pdf
34. Arctic Economic Council (2024) Arctic Mining Report 2024. https://arcticeconomiccouncil.com/wp-content/uploads/2024/10/aec-arctic-mining-report-2024-sample.pdf
35. The Arctic Institute (2023) The European Union and its Member States in the Arctic: Official Complementarity or Underlying Rivalry? https://www.thearcticinstitute.org/european-union-member-states-arctic-official-complementarity-underlying-rivalry/
36. Conley HA, Melino M (2020) America's Arctic moment: great power competition in the Arctic to 2050. Center for Strategic and International Studies (CSIS)
37. NATO (2022) NATO 2022 strategic concept. NATO Press Release https://www.nato.int/cps/en/natohq/topics_210907.htm
38. NATO. (2023). Vilnius Summit Communiqué. North Atlantic Treaty Organization.https://www.nato.int/cps/en/natohq/official_texts_217320.htm
39. Boulègue M Northern Front: NATO's security policy in the high north after finland and sweden's accession. Chatham House Research Paper, May 2023.https://www.chathamhouse.org
40. Byers M (2013) International law and the Arctic. Cambridge University Press
41. Erickson AS, Pincus R (2020) China's strategic intentions in the Arctic. The Maritime Executive
42. Pfeifer G, Szumski R (2023) Sabotage at sea: the Baltic Sea pipeline incident and the future of maritime critical infrastructure security. Eur Secur 32(2):264–281
43. Soldi G et al (2023) Monitoring of critical undersea infrastructures: the Nord stream and other recent case studies. IEEE Aerosp Electron Syst Mag 38(10):4–24. https://doi.org/10.1109/MAES.2023.3285075
44. NATO (2023) NATO launches undersea infrastructure security initiative. NATO Press Release https://www.nato.int/cps/en/natohq/news_213140.htm
45. Brzoska M, Fröhlich C (2016) Climate change, migration and violent conflict: vulnerabilities, pathways and adaptation strategies. Migr Dev 5(2):190–210
46. Biskaborn BK et al (2019) Permafrost is warming at a global scale. Nat Commun 10(1):264

Chapter 5
Drought Risk Management in the MENA Region: The Role of Earth Observations and Open-Source Tools

Jawad Al-Bakri and Muhammad Rasool Al-Kilani

Abstract Drought as a climate risk in the Middle East and North Africa (MENA) is related to disastrous implications on a wide range of aspects. This chapter presents an overview on drought risk management in the MENA region, the technical approaches, notable progress in the past 10 years, and the main restrictions for regional drought risk management. Within these contexts, this chapter discusses the role of earth observations and open-source tools in supporting drought mitigation efforts in the MENA region. Satellite reanalysis products which provide meteorological estimations help circumvent challenges related to limited meteorological data coverage, record gaps, and data accessibility. Open-source versions provide API capabilities and near-real time observations which could facilitate early warning systems. Increasingly popular sources like NASA's POWER platform, has been recommended with lesser restrictions on their use in drought analysis following validation studies in the MENA. Potential utilities include drought characterization in terms of intensity, duration, and frequency. Open-source reanalysis products also extend to provide proxy for impact data, such as biomass productivity, which are scarcer and comprise a much bigger challenge in drought vulnerability assessments. FAO's WaPOR platform is a prominent example, and institutionalization at national level in the MENA region has been supported by FAO which could significantly improve drought assessment capabilities. Gridded visualization of drought vulnerability based on both drought characterization and impact data is also available through the PRISM tool. Complementary tools to help identify more specific impacts include crop simulation and climate models. Capacity building and training is essential to fully utilize emerging tools due to the wide range of their applications.

J. Al-Bakri · M. R. Al-Kilani
School of Agriculture, University of Jordan, Amman, Jordan

M. R. Al-Kilani (✉)
School of Natural Resources Engineering and Management, German Jordanian University, Amman, Jordan
e-mail: rasoolkilani@live.com

© The Author(s), under exclusive license to Springer Nature Switzerland AG 2025
M. F. Buongiorno et al. (eds.), *The Challenge of Environmental Security in the Euro-Mediterranean Region*, NATO Science for Peace and Security Series C: Environmental Security, https://doi.org/10.1007/978-3-031-99646-7_5

Keywords Arid regions · Disaster preparedness · Reanalysis products · NASA POWER · WaPOR · PRISM

5.1 Drought in the MENA: Beyond Meteorology

Dominantly arid and semi-arid regions such as the Middle East and North Africa (MENA) are particularly vulnerable to drought impacts due to the fragile nature of their land resources [1]. The region has experienced record-breaking drought intensities in the past years in various locations including Jordan, Iraq, and Syria [2, 3]. The nature and scale of drought impacts are the defining characteristics, the soonest to identify is 'meteorological drought' at which a prolonged anomaly in meteorological measurements is indicated (Fig. 5.1), thus relevant drought protocols largely focus on this aspect [4–7]. Dry conditions can lead to an 'agricultural drought' by reducing soil water and impacting crops, and when reservoirs, streamflow, or groundwater is affected, it is categorized as a 'hydrological drought,' which typically lasts longer [4]. Drought impacts, influenced by vulnerability, can extend to socioeconomic and ecological aspects [4, 8].

Extreme droughts in the MENA have been shown to directly impact seasonal yields and increase crop failure and have been attributed to regional conflicts in the region [2, 9–11]. The Mediterranean basin and surrounding regions have also been identified as global hot spots for extreme compound drought events and heatwaves with severe implications on mortality and health [12–14]. The Arabian Peninsula in

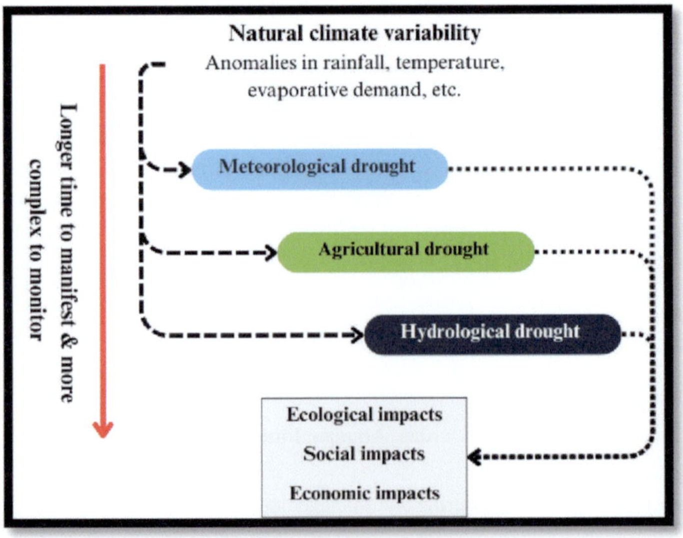

Fig. 5.1 Progression of drought leading to ecological and socioeconomic impacts

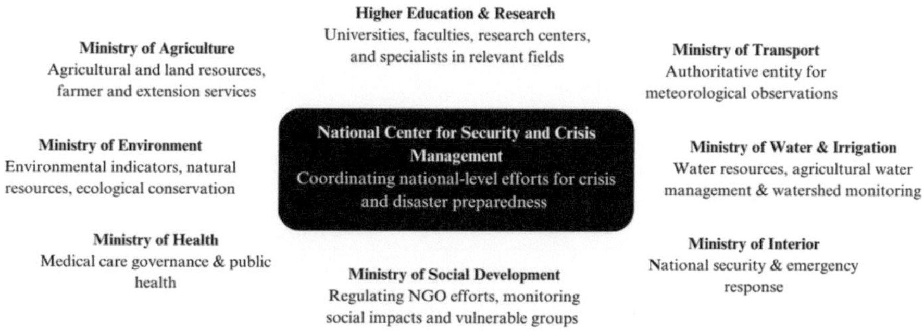

Fig. 5.2 Government institutions in Jordan involved in drought risk preparedness and mitigation and contributing to relevant national strategies

particular has experienced one of the fastest warming rates in the world throughout the past century [3]. The frequency, severity, and duration of droughts and other climate extremes have been projected to worsen under different climate scenarios in the MENA region [15–17]. Therefore, drought risk analysis is not simply an aspect of meteorology in the MENA, but rather a crucial task with implications on all aspects of life. This is reflected by the integration of various agencies in countries like Jordan for coordinating drought mitigation and relevant national strategies as part of the national efforts for disaster preparedness (Fig. 5.2).

5.2 Drought Risk Management in the MENA Region

Drought risk management essentially requires two types of data: measurements for drought characterization and impact data describing vulnerability. Timeseries of meteorological and moisture-related measurements are used to determine drought intensity, duration, and frequency [18–20]. Soil moisture is related to technical complexities, higher costs, and limited spatial coverage [21, 22], and so meteorological measurements are more widely adopted [6, 12]. The most common approach for drought characterization is through drought indices, which represent the deviation of a water variable (e.g. rainfall) from long term or 'normal' conditions (> 30 years of record) [4]. The most available gauged meteorological parameter globally is rainfall, and so the Standardized Precipitation Index (SPI) is the most widely recommended index [7]. A revised version was devised by Vicente-Serrano et al. [5] which also considers evapotranspiration (ET), named the Standardized Precipitation Evapotranspiration Index (SPEI). Categorical classifications are commonly assigned to SPI or SPEI values as shown in Table 5.1.

Figure 5.3 demonstrates drought conditions throughout the past 30 years in different locations in the MENA region. The data shows how the region is increasingly experiencing extreme droughts, which is consistent with predictions made in the past.

Table 5.1 Drought categories commonly assigned to SPI and SPEI values

SPI or SPEI value	Category
> 1	Wet conditions
0.99 to − 0.99	Near normal conditions
− 1 to − 1.49	Moderately dry
− 1.5 to − 1.99	Severely dry
≤ − 2	Extremely dry

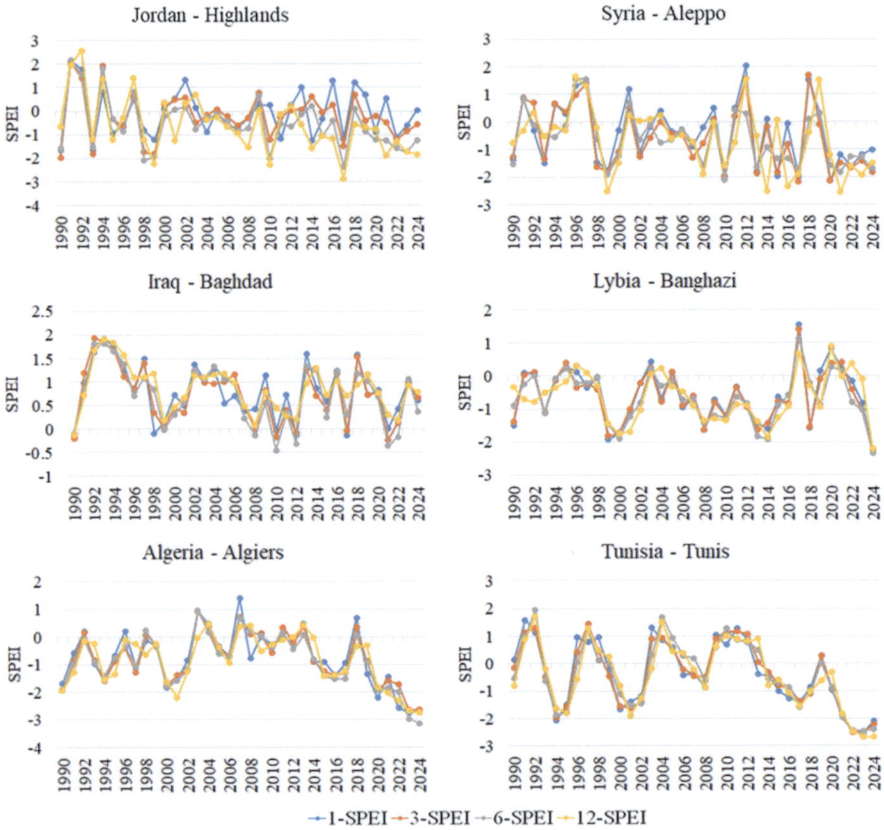

Fig. 5.3 Drought severity and frequency categorized by SPEI at 1-, 3-, 6-, and 12-month scales for different regions in the MENA (using the Thornthwaite model based on data from NOAA)

On the other hand, impact data reflects damages on an aspect effected by drought such as stream flow information, agricultural yields, or ground water levels to help characterize the 'Hazard' nature of drought [23–25]. Törnros and Menzel [26] explored the impacts of drought represented by SPI and SPEI on vegetation health in the Jordan River basin, reporting that the 6-month scale SPEI showed the highest correlation with soil moisture and vegetation health. Al-Bakri et al. [27] compared the

SPI and the Combined Drought Index (CDI) for assessing drought impacts on agriculture, reporting that crop yields responded most to the 6-month CDI (Nov-Apr) and that Jordanian highlands showed the highest vulnerability to drought impacts. These types of analysis elaborate which type of drought, timescale, period, and severity is particularly related to specific impacts. The Sustainable Development Goals (SDG) indicators specifically focus on this aspect. For example, SDG indicator 15.3 for monitoring land degradation focuses on drought affected lands, rather than drought severity itself. In the context of climate risk analysis, relying on the most widely available gauged measurement (rainfall) as a focal point for drought analysis could ignore the hazardous nature of drought. For example, increasing evidence suggests that SPEI is more reliable for reflecting impacts on agriculture [2, 24, 28].

5.3 Drought Assessment Restrictions and Strategies

For a proper framework of drought risk management, both meteorological measurements and impact data are essential. One of the major restrictions is the availability of meteorological measurements at sufficient spatial distributions, a challenge which has been reported in the MENA and other regions. Relevant issues included poor spatial distribution, faulty records, extreme topographic variability, and unreliability resulting from record gaps and technical problems [2, 29–32]. For example, some of the massive crop failures reported in Southern Jordan in 2021 and 2022 were not originally linked to drought due to limited accessibility and meteorological data coverage in those regions, specifically, the lack of real or near-real time measurements. Like many other countries, Jordan officially announces a drought emergency based on seasonal rainfall. The condition for announcing drought emergency in Jordan has commonly been whether seasonal rainfall dropped below 40% of its long-term average, which ignores the hazardous nature of drought. The availability of impact data is a much bigger issue; Hagenlocher et al. [33] found that more than half of the drought risk assessments did not consider the hazardous nature of drought, which could be related to impact data availability, especially in the agricultural sector. Labudová et al. [24] specifically pointed the availability of farm records as a restriction for drought impact assessment. Al-Kilani et al. [34] noted that farm records were largely unavailable, and when farmers did document, they were unreliable. Farmers tend to record information like “net profit” without documenting the percentage of the crop sold or market price. Drought risk assessments were mostly done at district or national levels due to relying on census data and lack of records at farm levels [24, 27, 34]. Another problem exists in the drought mitigation frameworks, which largely focus on monitoring drought indicators rather than specific analysis of the vulnerability of different sectors. The capacity for drought risk assessment can be enabled at the level of different institutions within a country, this is due to the unique impacts drought and compound drought events can inflict on different sectors. Cross-boundary and international collaboration is essential and commonly practiced, which takes different forms such as knowledge sharing and financing (Table 5.2).

Table 5.2 Examples on international and institutional collaborative drought mitigation efforts in the MENA region between 2010 and 2025

Country	Target institution and activity	Sponsor/ Implementation partner
Jordan	Establishment of The Agricultural Drought Unit at The National Agricultural Research Center	United Nations' World Food Program
Jordan	The University of Jordan—Project #983368 'Indicators of Drought and Desertification'	The North Atlantic Treaty Organization (NATO)—Science for Peace and Security Program
Jordan	Development of the national drought management plan at the Ministry of Water and Irrigation	United Nations' Development Program
Jordan	Development of drought action plans with national governments in collaboration with The International Water Management Institute (IWMI)	IWMI, funded by the USAID
Lebanon		
Morrocco		
Saudi Arabia	Formally joining the International Drought Resilience Alliance in 2024, a global coalition that mobilizes political, technical, and financial resources to improve preparedness to more severe droughts	International Drought Resilience Alliance

Additionally, drought vulnerability assessments can be at national level through technical committees who collect data and perform relevant evaluations in collaboration with various experts. In Jordan, this is done via the National Center for Security and Crisis management (Fig. 5.4).

The most notable challenges and gaps in drought mitigation and risk management in the MENA region can be summarized as follows [2, 33–36]:

(1) Availability of meteorological measurements with good coverage
(2) Availability of impact data at different scales and incorporating impact assessment
(3) Involvement of all relevant stakeholders in relevant policies
(4) Accessibility to data, information, and technical expertise
(5) Limited or lack of drought prediction in drought risk management plans
(6) Too focused on 'finding the right drought indicator' instead of drought hazards
(7) Lack of guidelines for assessing drought impacts and drought mitigation measures at the sector level.

Many of these challenges can be resolved through widely available and accessible data for drought characterization and impact assessment with good spatial coverage.

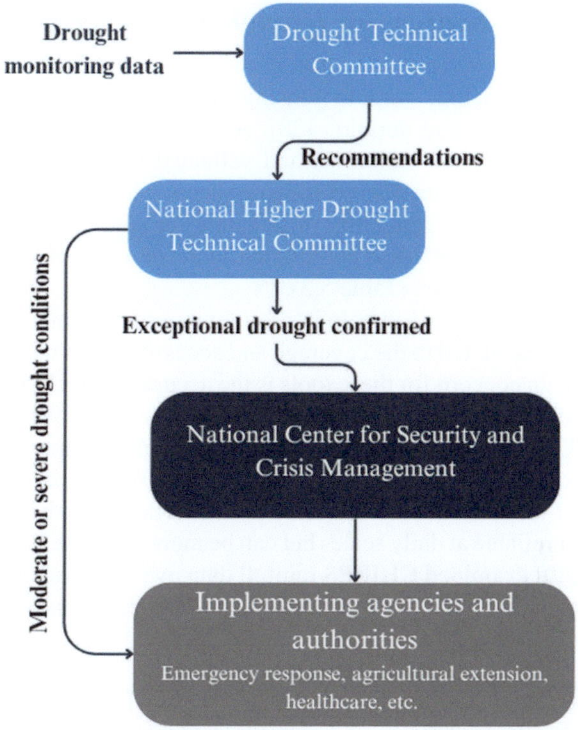

Fig. 5.4 The conceptual framework for the drought risk prevention strategy in Jordan

5.4 Application of Earth Observations and Open-Source Tools

Global coverage of different physical properties has become available through earth observations from satellites. Reanalysis products have taken this a step further by providing a wider range of gridded measurements based on advanced processing of satellite images (e.g. rainfall estimates from top-of-cloud properties). With the introduction of open-source satellite products via cloud computing, users can obtain near-real time measurements of a wide range of properties [37]. Furthermore, users can utilize it for Application Programming Interface (API) capabilities, allowing for automated data collection, processing, and different specialized analysis like customizing predictive models and developing early warning systems. Different sources are available for drought characterization, impact assessments, or even vulnerability analysis. This section provides an overview of such tools and resources, case study applications, and their utility in drought characterization in the context of the MENA region.

5.4.1 Sources for Drought Characterization

These include remote sensing-based data to help categorize drought intensity, frequency, and duration [38]. A very advantageous aspect of these sources is that they are often calibrated and validated through complex mathematical modelling based on ground-based measurements where ground data is highly available. This improves the reliability of this data for application in regions with limited ground data coverage. Most commonly, such sources include estimations of meteorological or relevant data (Table 5.3).

These tools can help in drought monitoring and characterization by avoiding the issues related to the coverage and accessibility to meteorological measurements. The main concern for these tools is the accuracy, which many researchers explored in the MENA region. Al-Sheriadeh and Al-Sharman [39] evaluated rainfall data from the TRMM and PERSIANN-CDR against ground measurements in Jordan reporting a root mean squared error between 9 and 20 mm/month for TRMM and between 11 and 41 mm/month for PERSIANN-CDR. The study also reported that both sources were unreliable at daily scales but can be more accurate at monthly scales. Mianabadi et al. [40] examined CHIRPS rainfall data over Iran, reporting strong correlations (> 0.8) at monthly and yearly scales, but poor correlations for daily data. The study analysed

Table 5.3 Open-source satellite products and their utility in drought monitoring

Satellite product	Product details	Utility in drought analysis
Global Precipitation Measurement (GPM)	NASA-JAXA satellite mission providing global precipitation data	Time series can be used to calculate rainfall-based indices (e.g. SPI)
Tropical Rainfall Measuring Mission (TRMM)	NASA-JAXA mission focused on tropical and subtropical rainfall	
Climate Hazards Group InfraRed Precipitation with Station Data (CHIRPS)	Global precipitation dataset based on satellite and ground stations data	
Modern-Era Retrospective Analysis for Research and Applications, Version 2 (MIRRA-2)	Global reanalysis dataset based on satellite and ground observations, provides temperature, humidity, wind speed, precipitation, pressure, and aerosols	Time series can be used to calculate rainfall-based indices (e.g. SPI) or complement ground measurements for other indices (e.g. SPEI)
ECMWF Reanalysis 5th Generation (ERA5)	High-resolution global atmospheric reanalysis by the European Centre for Medium-Range Weather Forecasts. Provides temperature, wind, humidity, precipitation, and pressure	

drought patterns over southern Iran using CHIRPS data and revealed that shorter-scale droughts are becoming shorter but with increased intensity, i.e. becoming more severe. Alsalal et al. [41] evaluated rainfall data from CHIRPS and the Climate Forecast System Reanalysis (CFSR) in Jordan, reporting that CHIRPS was more reliable at daily, monthly, and seasonal scales with higher correlation (0.76) compared to CFSR (0.65). Recently, several sources are grouped and presented via web apps and platforms. This provides users with improved accessibility, graphing capabilities, and reduced need for analysis. One of the prominent platforms for this purpose is The Prediction of Worldwide Energy Resources (POWER) project by NASA, which provides a wide range of variables from various satellite products as gridded data. Since its release, the platform has become popular for drought analysis in the MENA. Al-Kilani et al. [19] evaluated rainfall data from POWER for calculating SPI at different scales in Jordan compared to ground measurements, reporting that POWER based drought intensities were specifically reliable for detecting extreme droughts but not less intense or wet events. Kheyruri et al. [42] used POWER data for drought characterization by the SPI and SPEI over Iran, reporting that the highest drought severity occurred between 1999 and 2003, and that drought represented by SPI was more severe. Gebeyhu [43] applied bias correction to POWER data for drought characterization over the Nile River basin, improving R2 from 0.71 up to 0.96, the study found that only around 30% of the basin was under wet or near normal conditions, and mostly experienced drought conditions. Al-Kilani et al. [2] demonstrated how SPEI calculated from POWER reflected extreme drought conditions during massive crop failures in southern Jordan, despite rainfall being within the long-term normal average.

5.4.2 Impact Assessment Sources and Supplementary Tools

Some remote sensing-based products provide estimations related to the sectors impacted by drought, namely agriculture and vegetation. The most widely used indicator for this purpose is the Normalized Difference Vegetation Index (NDVI), which is used as proxy impact data to characterize vegetation health under different drought intensities [26, 27, 44]. The NDVI can be determined from Red and Near Infrared bands which are available in satellite missions such as Sentinel-2 and Landsat [45]. These can be downloaded from sources such as the United States Geological Survey (USGS) website, or by sources that directly provide NDVI such as:

- The National Center for Environmental Information (https://www.ncei.noaa.gov/)
- The Digital Drylands project (https://digital-dryads.eu/interactive-ndvi/)
- The online Streamlit App (https://ndvi-viewer.streamlit.app/)
- The Copernicus Browser (https://browser.dataspace.copernicus.eu/).

More sophisticated open-source tools have been developed, these provide biomass estimations for vegetation and agricultural lands. One of the most robust platforms for such purposes is FAO's portal to monitor Water Productivity through Open-access of Remotely sensed derived data (WaPOR). This tool provides a wide range of indicators to monitor biomass productivity at up to 30 m resolutions [46, 47]. This solves a major problem for drought impact assessment by providing a proxy for impact data. Initially, the portal covered the Middle East and Africa [48], but the recent version (WaPOR v3) provides global coverage. The tool has been used in the MENA region for various purposes; Hajirad et al. [49] examined water productivity and ET in the drought vulnerable Maroon-Jarahi basin in Iran. Seijger et al. [50] investigated factors influencing spatial variability in WaPOR data across farmlands in Ethiopia and Mozambique, finding that photosynthesis efficiency, local climate, and irrigation methods accounted for 82–94% of the observed variability. Al-Kilani et al. [34] explored WaPOR data as proxy for drought impact data in 2 rainfed farmlands in Jordan, reporting that water productivity layers from WaPOR showed strong correlation with drought intensity and that total biomass production was lower during drier periods. FAO has been promoting the institutionalization of WaPOR at national level in several MENA countries (Table 5.4).

These efforts will have a positive impact on drought assessment capabilities of national institutions in the MENA even when ground observations or survey data are scarce. Two or more open-source tools can be helpful for drought risk assessment, e.g. drought indices calculation using POWER with impact data from WaPOR. Alternatively, some platforms have been developed to accommodate these tasks in one tool, one primary example is the Platform for Realtime Impact and Situation Monitoring (PRISM) developed by the world food program. PRISM is a climate risk monitoring platform that support disaster risk reduction efforts by integrating geospatial data on hazards—including droughts, tropical storms, and floods—with socioeconomic vulnerability indicators (Fig. 5.5). Several national institutions in Jordan have received training on the utilization of this tool for drought risk management.

In addition to earth observation data, some of the most powerful tools to be used for drought impact assessment are climate and crop simulation models [51, 52]. These

Table 5.4 FAO efforts for institutionalizing WaPOR at national level in the MENA region

Period	Activity
2020–2022	Technical staff from The Ministries of Water and Agriculture in Egypt, Jordan, and Palestine are trained on incorporating WaPOR v2 within evaluation and monitoring activities in collaboration with IHE Delft and IWMI
2022–2023	Technical staff from The Ministries of Water and Agriculture in Jordan, and Palestine are trained on performing localized validation studies for WaPOR v2 in collaboration with IHE Delft and IWMI
2024–2025	Technical staff from Algeria, Tunisia, Sudan, Egypt, Jordan, and Palestine are trained on producing higher level data using PyWaPOR and using WaPOR v3 in monitoring and assessment in collaboration with IHE Delft
2025–2026	FAO and The University of Jordan collaborate to institutionalize WaPOR within the databases and protocols of the Ministry of Agriculture and Ministry of Water and Irrigation

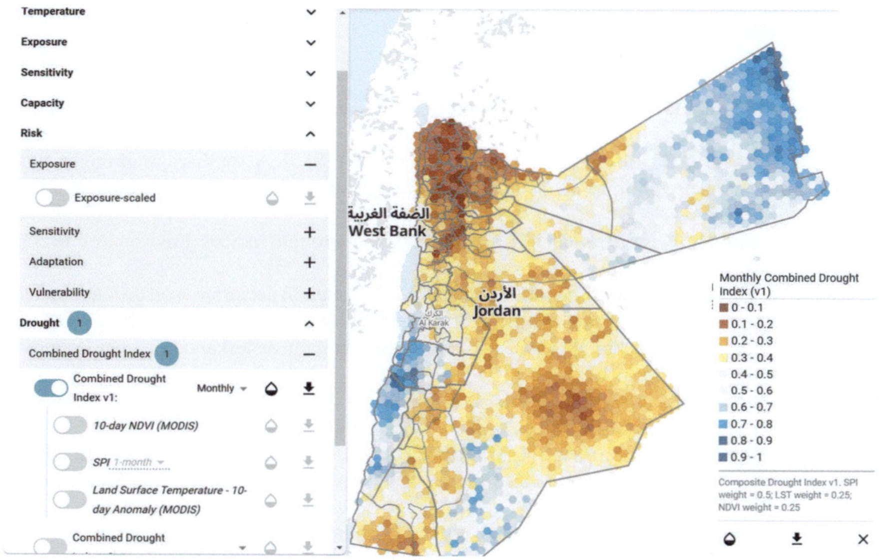

Fig. 5.5 Drought vulnerability assessment of Jordan using PRISM

can integrate simulated meteorological data to characterize drought conditions in the future and potential impacts on crop yields under simulated conditions. Al-Bakri et al. [53] integrated simulated projections from the NEX-GDDP project into the DSSAT crop simulation model to predict climate impacts on barely in Jordan. The study projected decreased precipitation and increased temperatures (2.0–5.4 °C) for 2030–2050 and 2080–2100 under RCP4.5 and RCP8.5. DSSAT simulations predicted up to 30% barley yield reductions between 2030–2050, and further up to 19% reduction by 2080–2100. These tools have proven effective in circumventing the gaps in data availability and accessibility. However, the full utilization of these tools for mitigating drought impacts in the MENA requires extensive capacity building. This is due to varying ways these tools can be applied for different sectors vulnerable to drought and climate variability. The integration of interdisciplinary expertise will also need non-technical users to understand the extent of interpreting remote sensing-based and simulated data in the context of drought risk management.

References

1. Al-Kilani MR (2024) Agricultural land measures for climate change adaptation in arid regions: can the farmers do it alone? jaa 82–93
2. Al-Kilani MR, Al-Bakri J, Rahbeh M et al (2025) Agricultural drought assessment in data-limited arid regions using opensource remotely sensed data: a case study from Jordan. Theor Appl Climatol 156

3. Hamed MM, Ullah S, Jasim Al-Hasani AA et al (2025) Rising threats of compound droughts and heatwaves in Mena: implications for population and agriculture under Paris agreement scenarios. Earth Syst Environ
4. Svoboda M, Fuchs B (2017) Handbook of drought indicators and indices. Drought Water Crises 155–208
5. Vicente-Serrano SM, Beguería S, López-Moreno JI (2010) A multiscalar drought index sensitive to global warming: the standardized precipitation evapotranspiration index. J Clim 23:1696–1718
6. Zargar A, Sadiq R, Naser B, Khan FI (2011) A review of drought indices. Environ Rev 19:333–349
7. Hayes M, Svoboda M, Wall N, Widhalm M (2011) The Lincoln declaration on drought indices: universal meteorological drought index recommended. Bull Am Meteor Soc 92:485–488
8. Yang X, Liao X, Di D, Shi W (2023) A Review of drought disturbance on socioeconomic development. Water 15:3912
9. Ide T, Lopez MR, Fröhlich C, Scheffran J (2020) Pathways to water conflict during drought in the MENA region. J Peace Res 58:568–582
10. Shahvari N, Khalilian S, Mosavi SH, Mortazavi SA (2019) Assessing climate change impacts on water resources and crop yield: a case study of Varamin plain basin. Environ Monit Assess, Iran, p 191
11. Kelley CP, Mohtadi S, Cane MA et al (2015) Climate change in the fertile crescent and implications of the recent Syrian drought. Proc Natl Acad Sci USA 112:3241–3246
12. Luque-García L, Bataineh S, Al-Bakri J et al (2024) The heat-mortality association in Jordan: effect modification by greenness, population density and urbanization level. Sci Total Environ 952:176010
13. Afroz M, Chen G, Anandhi A (2023) Drought-and heatwave-associated compound extremes: a review of hotspots, variables, parameters, drivers, impacts, and analysis frameworks. Front Earth Sci 10
14. Al-Delaimy WK (2020) Vulnerable populations and regions: Middle East as a case study. In: Health of people. Health of planet and our responsibility, pp 121–133
15. Zittis G, Almazroui M, Alpert P et al (2022) Climate change and weather extremes in the eastern Mediterranean and Middle East. Rev Geophys 60
16. Li H, Li Z, Chen Y et al (2021) Drylands face potential threat of robust drought in the CMIP6 SSPs scenarios. Environ Res Lett 16:114004
17. Tabari H, Willems P (2018) More prolonged droughts by the end of the century in the Middle East. Environ Res Lett 13:104005
18. Zhou Y, Zhou P, Jin J et al (2022) Drought identification based on Palmer drought severity index and return period analysis of drought characteristics in Huaibei Plain China. Environ Res 212:113163
19. Al-Kilani MR, Rahbeh M, Al-Bakri J et al (2021) Evaluation of remotely sensed precipitation estimates from the NASA POWER project for drought detection over Jordan. Earth Syst Environ 5:561–573
20. Oikonomou PD, Karavitis CA, Tsesmelis DE et al (2020) Drought characteristics assessment in Europe over the Past 50 Years. Water Resour Manage 34:4757–4772
21. Abdelal Q, Al-Kilani MR (2024) Versatile simplistic correction of T-higrow sensors for improved soil moisture measurement accuracy. Meas Sci Technol 35:095801
22. Al-Kilani MR, Abdelal Q, Al-Shishani G (2024) Are conductivity sensors useless for irrigators? Exploring measurement consistency around soil moisture thresholds relevant to different applications. Irrig Drain 74:1018–1030
23. Madruga de Brito M, Kuhlicke C, Marx A (2020) Near-real-time drought impact assessment: a text mining approach on the 2018/19 drought in Germany. Environ Res Lett 15:1040a9
24. Labudová L, Labuda M, Takáč J (2016) Comparison of SPI and SPEI applicability for drought impact assessment on crop production in the Danubian Lowland and the East Slovakian Lowland. Theor Appl Climatol 128:491–506

25. Naumann G, Spinoni J, Vogt JV, Barbosa P (2015) Assessment of drought damages and their uncertainties in Europe. Environ Res Lett 10:124013
26. Törnros T, Menzel L (2014) Addressing drought conditions under current and future climates in the Jordan River region. Hydrol Earth Syst Sci 18:305–318
27. Al-Bakri JT, Alnaimat et al (2019) Assessment of combined drought index and mapping of drought vulnerability in Jordan. Int J Eng Res
28. Vicente-Serrano SM, Beguería S, Lorenzo-Lacruz J et al (2012) Performance of drought indices for ecological, agricultural, and hydrological applications. Earth Interact 16:1–27
29. Boluwade A (2020) Spatial-temporal assessment of satellite-based rainfall estimates in different precipitation regimes in water-scarce and data-sparse regions. Atmosphere 11:901
30. Zhao Q, Chen Q, Jiao M et al (2018) The temporal-spatial characteristics of drought in the loess plateau using the remote-sensed TRMM precipitation data from 1998 to 2014. Remote Sens 10:838
31. Yassin MA, Alazba AA, Mattar MA (2016) Artificial neural networks versus gene expression programming for estimating reference evapotranspiration in arid climate. Agric Water Manag 163:110–124
32. Overeem A, Leijnse H, Uijlenhoet R (2013) Country-wide rainfall maps from cellular communication networks. Proc Natl Acad Sci USA 110:2741–2745
33. Hagenlocher M, Meza I, Anderson CC et al (2019) Drought vulnerability and risk assessments: state of the art, persistent gaps, and research agenda. Environ Res Lett 14:083002
34. Al-Kilani MR, Al-Bakri J, Rahbeh M et al (2024) Assessment of meteorological drought impacts on rainfed agriculture using remote sensing–derived biomass productivity. Environ Monit Assess 196
35. Bergaoui K, Fraj MB, Fragaszy S et al (2024) Development of a composite drought indicator for operational drought monitoring in the MENA region. Sci Rep 14
36. Fragaszy SR, Jedd T, Wall N et al (2020) Drought monitoring in the Middle East and North Africa (MENA) region: participatory engagement to inform early warning systems. Bull Am Meteor Soc 101:E1148–E1173
37. Farhan IA, Al-Bakri J (2019) Detection of a real time remote sensing indices and soil moisture for drought monitoring and assessment in Jordan. OJG 09:1048–1068
38. Al-Bakri JT, Al-Khreisat et al (2017) Assessment of remote sensing indices for drought monitoring in Jordan. Asian J Geo
39. Al-Sheriadeh M, Al-Sharman AR (2024) Evaluation of satellite rainfall estimates using PERSIANN-CDR and TRMM over three critical cells in Jordan. J Hydroinf 26:424–440
40. Mianabadi A, Salari K, Pourmohamad Y (2022) Drought monitoring using the long-term CHIRPS precipitation over Southeastern Iran. Appl Water Sci 12
41. Alsalal S, Mou LT et al (2023) Evaluation of CHIRPS and CFSR precipitation products over the Mujib Basin, Jordan. GEOGRAFIA 19
42. Kheyruri Y, Nikaein E, Sharafati A (2023) Spatial monitoring of meteorological drought characteristics based on the NASA POWER precipitation product over various regions of Iran. Environ Sci Pollut Res 30:43619–43640
43. Gebeyhu B (2024) Power data access viewer-based meteorological drought analysis and rainfall variability in the Nile river basin. Adv Meteorol 2024:1–12
44. Alahacoon N, Edirisinghe M (2022) A comprehensive assessment of remote sensing and traditional based drought monitoring indices at global and regional scale. Geomat Nat Haz Risk 13:762–799
45. Al-Bakri JT, D'Urso G, Batchelor C et al (2022) Remote sensing-based agricultural water accounting for the North Jordan Valley. Water 14:1198
46. WaPOR website. https://www.fao.org/inaction/remote-sensing-for-water-productivity/en/. Accessed Mar 2025
47. FAO (2024) WaPOR: the FAO portal to monitor water productivity through open access of remotely sensed derived data. FAO, Rome, Italy. https://data.apps.fao.org/wapor/?lang=en. Accessed Mar 2025

48. FAO (2020) WaPOR V2 quality assessment—technical report on the data quality of the WaPOR FAO database version 2. Rome, Italy, p 89. https://doi.org/10.4060/cb2208en
49. Hajirad I, Mohammadi S, Dehghanisanij H (2023) Determining the critical points of a basin from the point of view of water productivity and water consumption using the WaPOR database. In: The 7th international electronic conference on water sciences 86
50. Seijger C, Chukalla A, Bremer K et al (2023) Agronomic analysis of WaPOR applications: confirming conservative biomass water productivity in inherent and climatological variance of WaPOR data outputs. Agric Syst 211:103712
51. Alsalal S, Tan ML, Samat N et al (2024) Temperature and precipitation changes under CMIP6 projections in the Mujib Basin, Jordan. Theor Appl Climatol 155:7703–7720
52. Al-Bakri JT, D'Urso G, Calera A et al (2022) Remote sensing for agricultural water management in Jordan. Remote Sens 15:235
53. Al-Bakri J, Farhan I, Al-Qinna M et al (2021) Assessment of climate changes and their impact on barley yield in Mediterranean environment using NEX-GDDP downscaled GCMs and DSSAT. Earth Syst Environ 5:751–766

Chapter 6
Assessing Critical Infrastructure Exposure to Sea Level Rise and Land Subsidence Under Climate Change: The Cases of Venice and Fiumicino Airports and Marghera Harbour (Italy)

Marco Anzidei, Tommaso Alberti, Daniele Trippanera, Cristiano Tolomei, and Massimo Chiappini

Abstract Climate change-induced sea level rise threatens coastal infrastructures, particularly in densely populated and economically significant regions such as the Mediterranean. This study examines the exposure of three Italian coastal areas where ports, airports, buildings, tourist infrastructures, and communication networks are exposed to sea level rise. We focus on the major international airports of Venice Marco Polo and Rome Fiumicino Leonardo Da Vinci, located close to sea level, and the important commercial harbor of Porto Marghera, near Venice. These hubs are essential for global connectivity and economy but are becoming increasingly vulnerable to coastal flooding, storm surges, and extreme weather events due to ongoing global warming. Using recent projections from the Intergovernmental Panel on Climate Change and high-resolution Digital Elevation Models, we assess the potential inundation risks up to 2150 AD under different sea level rise scenarios, considering the high-emission Shared Socioeconomic Pathway SSP5-8.5. By analyzing these critical transport nodes' projected impacts and adaptive capacity, we highlight the urgent need for coordinated action to mitigate climate risks in Mediterranean coastal zones. Our study underscores the importance of integrating climate resilience into long-term infrastructure planning to safeguard people, economy, social stability, and environmental areas in the investigated region.

Keywords Sea level rise · Climate change · Land subsidence · DEM · Coastal hazard · Mediterranean sea

---

M. Anzidei · T. Alberti (✉) · D. Trippanera · C. Tolomei · M. Chiappini
National Institute of Geophysics and Volcanology (INGV), Rome, Italy
e-mail: tommaso.alberti@ingv.it

© The Author(s), under exclusive license to Springer Nature Switzerland AG 2025
M. F. Buongiorno et al. (eds.), *The Challenge of Environmental Security in the Euro-Mediterranean Region*, NATO Science for Peace and Security Series C: Environmental Security, https://doi.org/10.1007/978-3-031-99646-7_6

6.1 Introduction

Anthropogenic climate change is one of the most pressing global challenges of the twenty-first century, which is still causing multiple impacts on people, infrastructures, and the environment. Among the various effects of the ongoing global changes, sea level rise (SLR) is one of the most critical consequences threatening about one billion people living along the worldwide coastal regions. The Mediterranean area, characterized by densely populated coastlines, historical heritage sites, and vital economic hubs and coastal infrastructures, is particularly vulnerable to the impacts of rising sea. Over the past century, global sea levels have continuously accelerated due to the thermal expansion of the oceans and the melting of glaciers and ice sheets. Recent estimates shown in the AR6 Report of the Intergovernmental Panel on Climate Change (IPCC, www.ipcc.ch) indicate that the Mediterranean Sea has risen by approximately 1.4 mm per year over the past century, with an increased rate of around 3 mm/yr in the last three decades, now accelerating at about 4 mm/yr [1, 2]. Regional climate dynamics, vertical land movements (VLM), and changes in ocean circulation influence local variations in SLR. In particular, land subsidence (LS) can even locally triple the effects of the eustatic component alone [3]. The consequences of these changes are already being observed in many coastal areas, where erosion, coastal retreat, saltwater intrusion, and more frequent flooding events are affecting infrastructures, ecosystems, and coastal communities [2].

Storm surges and extreme weather events are yet exacerbating the risks associated with SLR on coastal infrastructures. An analysis of network vulnerability in the EU Countries estimated that accessibility indicators at the regional level showed up to a 26% decrease [4]. Since the last centuries, coastal cities such as Venice [5], Alexandria [6] and Valencia [7], besides small islands such as Lipari in the Aeolian archipelago [8], have experienced land subsidence, severe SLR and floodings, resulting in economic losses, disruptions to critical infrastructure and threats to cultural heritage. Studies indicate that extreme sea level events that were rare until a few decades ago (generally occurring once per century) could become annual occurrences by 2100 and beyond under high-emission scenarios [9]. Additionally, Mediterranean Sea surface temperatures have increased by approximately 1.5 °C since the late nineteenth century and are projected to rise by up to 3 °C by 2100, further intensifying extreme weather patterns [10]. Future projections indicate that, under high greenhouse gas emission scenarios, sea levels could rise by up to 1.1 m by the end of the century [1, 2]. Such an increase would significantly amplify the risks faced by low-lying coastal areas, endangering about one billion people and critical assets on a global scale. The Mediterranean basin, with its 46,000 km of coastlines along which are located large cities and small villages, intricate geopolitical assets and economic interdependencies, needs, therefore, to prioritize coordinated efforts to mitigate the increasing climate risks.

In this complex environmental framework, the risk associated with SLR to harbors, airports, and touristic infrastructures is relevant. Previous studies have identified 269 global airports at risk of flooding, which could grow up to 572 by 2100 [11], while in-depth analysis revealed that the impacts on ports will also be relevant. About 25% more cargo can be affected by extreme water levels by the end of the century, according to the high emission scenario, but depending on their geographical location and already existing protection measures [12].

Based on the above scenarios, in this work we have investigated the exposure of critical infrastructure to SLR, focusing on the two major international coastal airports of Venice Marco Polo (VCE) and Fiumicino Leonardo Da Vinci (FCO) and the coast near Rome between Ostia, Fiumicino and Passo Oscuro, and the industrial area of Porto Marghera, near Venice. These hubs are essential for regional connectivity and economic stability but are increasingly vulnerable to coastal flooding, storm surges, and extreme weather events. By using the latest AR6 projections from the IPCC and high-resolution geodetic data from ground GNSS networks and space based InSAR sensors, we estimated the relative sea level rise (RSLR) from the current rates of land subsidence to include in the analysis of the potential inundation risks under the worst SLR scenario corresponding to high-emissions for the Shared Socioeconomic Pathways SSP5-8.5 [2].

6.2 Materials and Methods

To assess the effects of RSLR by 2150 and create multi-temporal flooding maps, we used a multidisciplinary approach [5] combining: (i) GNSS data collected in the period 1996–2023, to measure ground velocity; (ii) SAR satellite data acquired in the time spanning June 2017–October 2023, to estimate the spatial rates of VLM; (iii) LiDAR-based DEMs from CO.RI.LA and the Italian Ministry of the Environment to project flooding extents; and (iv) IPCC AR6 sea level projections for SSP5-8.5 in the Mediterranean basin.

6.2.1 Geodetic Analysis

VLM rates were derived by integrating GNSS and InSAR data in the investigated sites (Fig. 6.1). For the GNSS data, we used the results of the Euro-Mediterranean velocity solution (e.g., ASI, GeoDAF, EPN, OGS-FredNet, and INGV RING) processed by the GAMIT/GLOBK software by the INGV geodetic data analysis center [13]. The GNSS velocity estimates were referenced to the IGb14 frame and corrected for station offsets and non-linear deformations. Only stations with a minimum 4.5-year time span were considered to ensure reliable velocity estimates (for further details, see [13]). Multi-temporal InSAR (MT-InSAR) was used to analyze ground motion over time. This technique, based on a large dataset of SAR images, improves accuracy compared to standard InSAR methods. We applied the Enhanced PS (EPS) approach, which combines Permanent Scatterers (PS) and Distributed Scatterers (DS), improving spatial coverage and measurement density. The resulting ground deformation maps achieve an accuracy of approximately $\pm$ 1.5 mm/yr. Sentinel-1

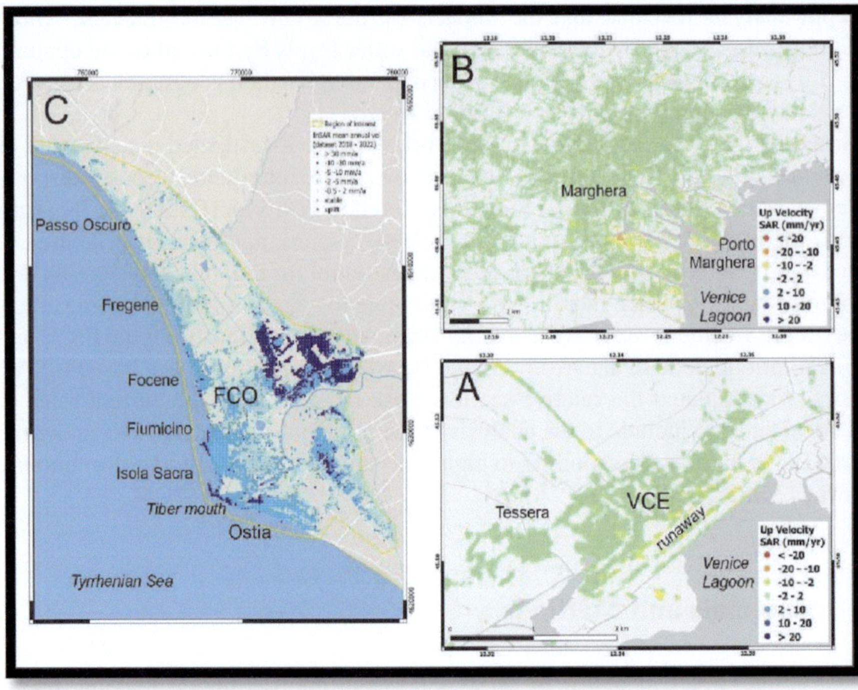

Fig. 6.1 EPS solution of the ground vertical velocity map for **a** the Venice airport Marco Polo and **b** Porto Marghera for the period 22 June 2017–25 October 2023. For **c** the coast from Ostia to Passo Oscuro, across FCO airport, the VLM has been obtained by using the ground motion service of Copernicus SAR products. InSAR velocities have been calibrated on the local GNSS stations

SAR images from 2017 to 2023 were also analyzed. The data were corrected for topographic effects using the ALOS-World 30 m DEM, and atmospheric noise was filtered out. The final velocity map was obtained by geocoding the results. Since our goal is to focus on the vertical component of land movements, the weak horizontal movements were neglected in the analysis [13]. We remark that InSAR velocities are calibrated by the nearby GNSS ground stations for the three investigated areas (Table 6.1).

6.2.2 IPCC Projections

We use the latest climatic projections based on the Sixth Assessment Report of the IPCC, publicly available at https://zenodo.org/record/6382554. They consist of median sea-level values and standard errors on a worldwide grid obtained by summing up the contributions of geophysical sources that are driving the long-term sea-level changes. Projections are provided for five Shared Socioeconomic Pathways (SSPs), outlining different future scenarios based on the expected socioeconomic

Table 6.1 RSLR and its components for the three areas (values are in m). SLR: sea level rise by IPCC for SSP 8.5 at 2150; VLM: vertical land motion by InSAR by considering 2023 as a benchmark year and linearly extrapolated for 2150; RSLR: that is the sum of SLR + VLM. The mean rate of areal land subsidence from InSAR data falls in the range of 1–3 mm/yr (Marghera <1 mm/yr; Venice Airport 1.2 mm/yr and Fiumicino 3.14 mm/yr)

Area	SLR IPCC (m)			VLM (m)			RSLR (m)		
	min	mean	max	min	mean	max	min	mean	max
Marghera	0.74	**1.16**	1.77	−0.06	**−0.11**	−0.13	0.80	**1.27**	1.90
Venice airport	0.74	**1.16**	1.77	−0.10	**−0.15**	−0.19	0.84	**1.31**	1.96
Fiumicino	0.95	**1.28**	1.82	−0.20	**−0.40**	−0.44	1.15	**1.68**	2.26

RSLR and its components for the three areas (values are in m). SLR: Sea Level Rise by IPCC for SSP 8.5 at 2150; VLM: Vertical Land Motion by InSAR by considering 2023 as a benchmark year and linearly extrapolated for 2150; RSLR: that is the sum of SLR + VLM. The mean rate of areal land subsidence from InSAR data falls in the range of 1-3 mm/yr (Marghera 1 mm/yr; Venice Airport 1.2 mm/yr and Fiumicino 3.14 mm/yr).

conditions and resulting temperature projections up to 2150: SSP1 (Sustainability—Taking the Green Road) envisions a world focused on sustainable development, leading to low emissions and limiting warming to around 1.5–2 °C; SSP2 (Middle of the Road) follows a trajectory of moderate progress and challenges, resulting in 2–3 °C of warming; SSP3 (Regional Rivalry—A Rocky Road) depicts a fragmented world with regional conflicts and slow economic growth, driving emissions and warming of 3—4 °C. SSP4 (Inequality—A Road Divided) highlights deep inequalities, where high emissions from developed regions push warming to 3—4 °C, and SSP5 (Fossil-fueled Development—Taking the Highway) assumes rapid economic growth driven by fossil fuels, leading to the highest warming of 4—5 °C.

The projected SLR is then estimated by combining the SLR rate at the grid points closest to the interested location (e.g., VCE, FCO, the coast between Ostia and Passo Oscuro and Porto Marghera), as provided by the IPCC projections, with the rate of VLM inferred from the geodetic analysis over the corresponding region. In detail, the measured rate of VLM, which includes GIA and tectonic components of both natural and anthropogenic origin, replaces the modeled GIA contribution used in the original IPCC AR6 projections. Assuming a constant rate of VLM until 2150 AD, in the absence of unpredictable tectonic events, the total amount of the VLM component for each area is strictly related to the time spanning from the reference year (or benchmarking year) and the projection year. For this study, the epoch 2023 AD was selected as the benchmark year from the IPCC report AR6.

However, the topographic dataset is based on a LiDAR (Light Detection And Ranging) survey carried out between 2008 and 2010, projected in the ITALGEO 2005 geoid [14]. Therefore, we have referred to this epoch our revised RSLR levels and the extension of the expected flooding areas.

6.3 Results

In the following section, we show the results of the integrated data analysis, with the projected RSLR up to 2150 for the SSP5-8.5 climatic scenario of the IPCC AR6 Report. In addition, the high-resolution multi-temporal maps of flooding scenarios for 2050, 2100, and 2150 AD for the three investigated areas are shown.

The Venice Marco Polo Airport

The Venice Marco Polo Airport (VCE) is located along the coast of the Venice Lagoon, and it represents the main touristic hub for Venice, with about 13 million passengers in 2024 (+ 20% of 2023), about 110,000 flights (+ 2.8% of 2023), 52 air companies, serving more than 90 destinations in 36 countries (data from the EU Interreg Project ReSites, 2017).

Figure 6.1 shows the results of our analysis for VCE, whose infrastructures and runaways are highly exposed to RSLR, being placed at very low elevation above the current mean SL. Given the mean LS at about 1.2 mm/yr for this area, the expected RSLR for the SSP5-8.5 scenario is just relevant in 2050 when the projected SLR will rise by about 26 cm, while in 2150, the expected mean SL could rise up to 1.31 m (referred to the benchmark year 2023). For this scenario, the potential flooding area is about 14 km^2, corresponding to about 74% of the surface surrounding the runaways. In the case of a maximum height water level (MHWL) event, as occurred in 1966 and 2019 in the Venice Lagoon (https://www.comune.venezia.it/it/content/centro-previsioni-e-segnalazioni-maree), the expected SL would rise up to 3.25 m (benchmark 2023), thus flooding about 92% of the area. In this extreme MHWL scenario, the airport runways will be completely submerged, and more than 800 buildings with a surface area greater than 20 m^2 that fall within the projected inundation area will also be affected (Fig. 6.2).

The harbor of Porto Marghera and Mestre

Porto Marghera is one of the largest coastal industrial areas in Europe, with over 220,000 hectares of surface area. It is an important economic hub for the entire northeastern sector of Italy due to its size, production, and employment characteristics. The economic volume of this infrastructure corresponds to approximately to 6.4 billion of estimated value of infrastructures (technological and industrial networks, IT networks, emergency management systems), employing about 11,000 workers in about 900 companies, mobilizing 25 million tons of goods on 26 industrial and commercial terminals, 163 berths with 1.5 million passengers distributed on 8 cruise terminals, for about 27 billion of estimated yearly turnover. The area of Marghera harbor has been affected by anthropogenic land subsidence during 1950–1970 due to the intensive industrialization and depletion of groundwater withdrawal that started with the onset of the industrial installations after 1930. During this period, this area experienced a total subsidence of 14 cm, which caused problems for the entire zone. In the recent decade, the mean LS has reduced at about 1 mm/yr, which is near to the natural geological one [15]. Therefore, in the absence of any protection systems, such as the MoSE [5, 16, 17], the expected RSLR at 2150 (considering 2023 as the benchmark) for the SSP.8.5 scenario is 1.27 m (slightly less than for nearby Marco

Fig. 6.2 RSLR scenario for the Venice Airport with the expected extension of flooding in 2050, 2100, and 2150 for the SSP5-8.5 scenario (see color scale). The maximum expected RSLR is 1.31 m in 2150 (benchmark year 2023) with a loss of land of about 14 km^2 estimated on the DSM (at 2008 AD). In the red scale, the SSP5-8.5 scenario up to 2150. The light-yellow area is the MHWL in case of a 1966 or 2019 type event at 2150 for the SSP5-8.5 scenario

Polo Airport). However, with this rise, about 20 km^2 of land is expected to be flooded, affecting more than 5800 buildings with a surface area greater than 20 m^2 and infrastructures, falling within the exposed area (enclosed in the purple line of Fig. 6.2), where more than 17,000 people are living.

In the case of MHWL in 2150, SL could rise up to 3.21 m in the SSP8.5 scenarios, corresponding to about 78 km^2 of flooded land. Flooding extension would also affect the densely inhabited area of Mestre and the local infrastructure network (about 44% of the area, as shown in Fig. 6.3).

The coast near Fiumicino

The investigated area consists of about 25 km of coastal tract between Marina di San Nicola in the north and Lido di Ostia in the south, across the Tiber River mouth with the village of Fiumicino (Fig. 6.4). The flat morphology of this coastal area with typical Mediterranean vegetation and small reliefs made up of dunes largely covered by buildings, is shaped by the evolution of the Tiber River mouth and the recent coastal erosion. In particular, the intense coastal retreat that started in the last two to three decades, and during relevant storms when many beach resorts have repeatedly suffered severe damages, finding themselves increasingly exposed to the action of the sea. The mean altitude of the investigated zone is about 5 m, although many inner areas are placed even less than 1 m above sea level, making it highly exposed to SLR. The coast is occupied by many tourist and economic installations, such as approximately 150 bathing establishments, harbors, and fishing centers. Fiumicino

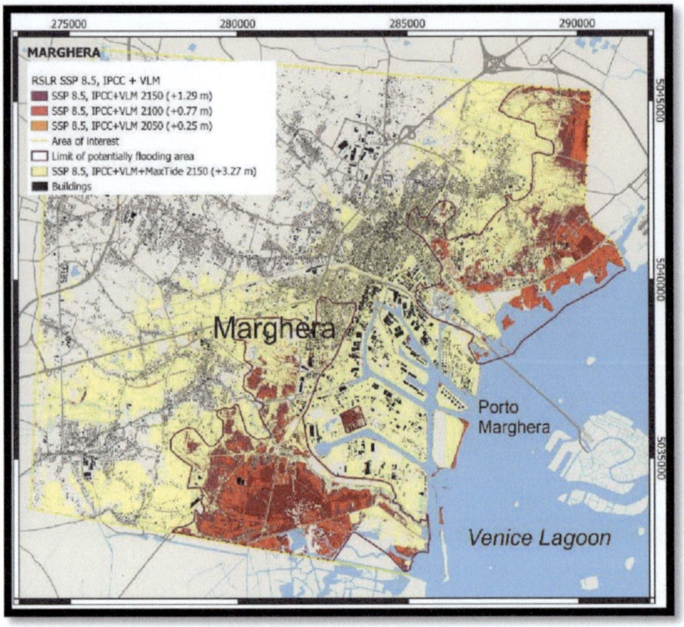

Fig. 6.3 RSLR scenario for Porto Marghera and Mestre, near Venice. See the color scale for the expected extension of flooding for 2050, 2100, and 2150 for the SSP5-8.5 emission scenario. The maximum expected RSLR is 1.27 m in 2150 (benchmark year 2023) with a loss of land of about 20 km^2 (estimated over a DSM at 2008 AD). The RSLR indicated in the Figure is estimated from LiDAR data. The red area is the SSP5-8.5 scenario up to 2150. The light-yellow area represents the MHWL, which is the combination of the maximum storm surge level at 2150 for the SSP5-8.5 scenario. The purple line is the area potentially directly or indirectly affected by RSLR

is one of the main fishing centers of the Tyrrhenian Sea and of great economic importance. And is the subject of important tourism projects. A new commercial port is planned in downtown Fiumicino, and another one for cruise ships in Isola Sacra. In addition, in Ostia and along the Tiber mouth, there are several tourist docks and shipyards. The area is interested in a dense and busy road network, especially during the summer season, with the railway lines Fiumicino Airport—Rome and the more coastal one Rome—Civitavecchia.

The main transport infrastructure in this area is the "Leonardo da Vinci" International Airport, which is located very close to the coastline of the Tyrrhenian Sea and at an altitude even less than 1 m above sea level (a.s.l). It is the largest in Italy and the third in Europe with a surface area of 29 km^2, as well as the busiest in the country. It also has 5 accommodation facilities. With 49.2 million passengers in 2024, it is the first Italian airport used by over 100 air companies connecting about 200 destinations and more than 70 counties. In addition, it is the second Italian airport for cargo traffic (https://assaeroporti.com/serie-storiche/).

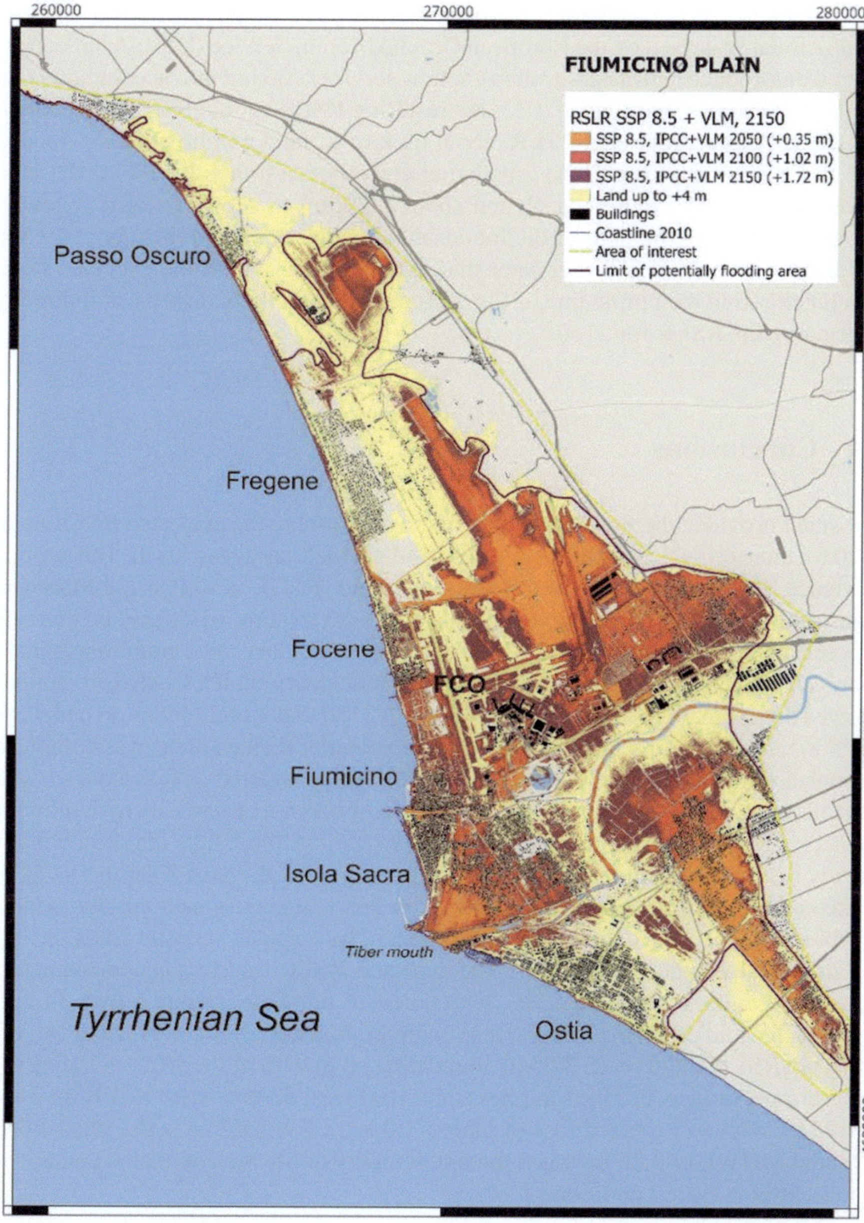

Fig. 6.4 The coast from Passo Oscuro, Fiumicino airport (FCO), and Ostia, near Rome. See the color scale for the expected extension of flooding in 2050, 2100, and 2150 for the SSP5-8.5 emission scenario. The maximum expected RSLR is 1.68 m in 2150 (benchmark year 2023) with a loss of land of 62 km^2 (calculated over a DSM of the year 2008). The RSLR is estimated from LiDAR surveys. In red, the SSP 5-8.5 scenario is up to 2150. The light yellow area represents the actual land below or equal to 4 m a.s.l. The purple line represents the possible area that will be directly or indirectly affected by RSLR. Note the impact of flooding on the local infrastructures of the Leonardo Da Vinci Airport, the villages of Fiumicino, Focene, and Ostia

The mean VLM rate of the area has been estimated at 3.14 mm/yr from the analysis of InSAR data released by the European Ground Motion Service (https://land.copernicus.eu/en/products/european-ground-motion-service). Taking into account this rate of LS and the benchmark year 2023, the resulting RSLR for the SSP5-8.5 scenario at 2150 is 1.68 m. With this RSLR, about 62 km^2 of land will be possibly flooded (Fig. 6.4). The area exposed to inundation corresponds to about 51% of the land placed below 4 m of altitude a.s.l. and about 122 km^2, as shown in light yellow in Fig. 6.3. Following the path of the inundated area, we estimated that about 19.500 buildings with a surface area greater than 20 m^2 falling within the expected flood area (enclosed in the purple line of Fig. 6.3) can be potentially directly or indirectly affected by the RSLR for 2150.

6.4 Conclusions

Our study provided the first assessment of the exposure to the projected RSLR up to 2150 for three critical infrastructures located in the Mediterranean basin. The airports of Venice Marco Polo (VCE) and Rome Fiumicino (FCO), as well as the Marghera industrial site and the coast between Ostia and Passo Oscuro, will likely be severely exposed to permanent flooding by the end of this century. The multidisciplinary approach that integrates GNSS and InSAR geodetic data with IPCC AR6 projections allowed the estimation of RSLR scenarios up to 2150 under high-emission pathways SSP5-8.5. We underline that the extent of the potential flooded areas was computed through a standard passive "bathtub" approach. We considered as flooded each area that falls below a defined water level and, in addition, is not necessarily hydraulically connected to the sea. The proposed flooding scenarios do not take into account any already existent adaptation or protection system, such as the MoSE for the VCE, or that could be realized before 2150. Although this is a simple method, this widely accepted approach can define the flooding hazard for a specific coastal area [18, 19].

For Venice Marco Polo Airport, the expected RSLR by 2150 is approximately 1.31 m above the current mean SL, with potential flooding extending over 14 km^2 (74% of the studied area). Under a maximum high-water level (MHWL) event, the projected RSLR could reach 3.25 m, inundating up to 92% of the area, including the entire airport surface. For the Marghera industrial area, the estimated RSLR for 2150 is 1.27 m, with an expected loss of land of 20 km^2. Even worse is the impact of a potential MHWL of 3.25 m, when the municipality of Mestre could also be flooded in this extreme event condition.

For the coast between Ostia, Passo Oscuro, and the Leonardo Da Vinci Airport, the expected RSLR by 2150 is up to 1.68 m for the SSP5-8.5 climatic scenario. In this case, the airport runaways will be partially flooded, as well as 18,000 buildings placed between the coastline and the maximum inland flooding extension. Our results highlight the urgent need for proactive planning and mitigation strategies to safeguard critical infrastructure in the Italian region against rising sea levels. The projected inundation risks for the investigated areas highlight the substantial threats posed by

SLR, exacerbated by land subsidence and extreme weather events. Given the essential role these considered hubs play in regional and international connectivity and the economic benefits that come from seaside and nautical tourism activities, disruptions due to flooding could have cascading effects on economic stability, tourism, emergency response capabilities and people safety. From a broader perspective, these results emphasize the necessity for long-term adaptation strategies, including the reinforcement of coastal defenses, revision of urban planning policies and integration of climate risk assessments into infrastructure development. The high-resolution geodetic approach employed in this study demonstrates the value of combining multiple data sources to enhance the accuracy of risk assessments, providing a model that can be extended to other vulnerable coastal locations. Beyond aviation infrastructure, the implications of rising sea levels extend to ports, energy facilities, and transportation networks, all of which are fundamental to economic resilience and security. Effective adaptation will require coordinated international efforts, integrating scientific research with policymaking and investment in climate-resilient infrastructure. Additionally, these findings reinforce the need to prioritize emissions reductions to mitigate worst-case scenarios, as adaptation measures alone may be insufficient to counteract the projected impacts under high-emission pathways. Moving forward, continued instrumental monitoring of VLM and SL trends, coupled with refined climate projections, will be crucial in informing timely interventions. Future research should also explore the interplay between RSLR, storm surges, and socio-economic vulnerabilities to develop more comprehensive risk and related economic frameworks. In this context, interdisciplinary collaboration between climate scientists, engineers, urban planners, and policymakers will be essential to ensuring sustainable and resilient coastal development to face the multiple impacts of SLR, LS, and land submersion triggered by the ongoing climate change.

Acknowledgements We are thankful to the Italian Ministry of the University and Research that funded this research under the project GAIA, contract no. PRIN2022_PE10_2022ZSMRXJ02. This research reflects the methodology applied in the SAVEMEDCOASTS (Agreement number ECHO/SUB/2016/742473/PREV16) and SAVEMEDCOASTS-2 (Project number 874398) EU Projects (www.savemedcoasts.eu; www.savemedcoasts2.eu), both funded by DG ECHO. LiDAR data were made available by CO.RI.LA, https://www.corila.it/ (accessed on 20 February 2025), and the Italian Ministry of the Environment to generate the DEMs at 1 m resolution in the ItalGeo 2005 geoid model.

Competing Interests The authors have no conflicts of interest to declare that are relevant to the content of this chapter.

References

1. Oppenheimer M et al (2019) Sea level rise and implications for low lying islands, coasts and communities. In: Pörtner H-O, Roberts DC, Masson-Delmotte V, Zhai P, Tignor M, Poloczanska E, Mintenbeck K, Alegría A, Nicolai M, Okem A (eds) IPCC special report on the ocean and cryosphere in a changing climate. IPCC, Geneva, p 126

2. Fox-Kemper B, Hewitt HT, Xiao C, Aðalgeirsdóttir G, Drijfhout SS, Edwards TL, Golledge NR, Hemer M, Kopp RE, Krinner G et al (2021) Masson-Delmotte V, Zhai P, Pirani A (eds) Ocean, cryosphere, and sea level change. In: Climate change: the physical science basis. Contribution of working group I to the sixth assessment report of the intergovernmental panel on climate change. Cambridge, UK
3. Vecchio A, Anzidei M, Serpelloni E (2023) Sea level rise projections up to 2150 in the northern Mediterranean coasts. Environ Res Lett 19:014050
4. Demirel H, Kompil M, Nemry F (2015) A framework to analyze the vulnerability of European road networks due to sea-level rise (SLR) and sea storm surges. Transp Res Part A: Policy Pract 81:62–76. https://doi.org/10.1016/j.tra.2015.05.002
5. Anzidei M, Tolomei C, Trippanera D, Alberti T, Bosman A, Brunori CA, Serpelloni E, Vecchio A, Falciano A, Deli G (2025) Multi-temporal relative sea level rise scenarios up to 2150 for the Venice Lagoon (Italy). Remote Sens 17:820. https://doi.org/10.3390/rs17050820
6. Stanley J-D, Clemente PL (2017) Increased land subsidence and sea-level rise are submerging Egypt's Nile delta coastal margin. GSA Today 27:4–11
7. Vousdoukas MI, Mentaschi L, Voukouvalas E et al (2018) Global probabilistic projections of extreme sea levels show intensification of coastal flood hazard. Nat Commun 9:2360. https://doi.org/10.1038/s41467-018-04692-w
8. Anzidei M, Bosman A, Carluccio R, Casalbore D, Caracciolo FD, Esposito A, Nicolosi I, Pietrantonio G, Vecchio A, Carmisciano C et al (2017) Flooding scenarios due to land subsidence and sea-level rise: a case study for Lipari Island (Italy). Terra Nova 29:44–51
9. IPCC (2022) Climate change 2022: impacts, adaptation, and vulnerability. contribution of working group II to the sixth assessment report of the intergovernmental panel on climate change. Cambridge University Press. https://doi.org/10.1017/9781009325844.021
10. Seneviratne SI et al (2021) Weather and climate extreme events in a changing climate. In: Masson-Delmotte V et al (eds) Climate change 2021: the physical science basis. Contribution of working group I to the sixth assessment report of the intergovernmental panel on climate change. Cambridge University Press, pp 1513–1766. https://doi.org/10.1017/9781009157896.013
11. Yesudian AN, Dawson RJ (2021) Global analysis of sea level rise risk to airports. Clim Risk Manag 31:100266. https://doi.org/10.1016/j.crm.2020.100266
12. Christodoulou A, Christidis P, Demirel H (2019) Sea-level rise in ports: a wider focus on impacts. Marit Econ Logist 21:482–496. https://doi.org/10.1057/s41278-018-0114-z
13. Serpelloni E, Cavaliere A, Martelli L, Pintori F, Anderlini L, Borghi A, Randazzo D, Bruni S, Devoti R, Perfetti P et al (2022) Surface velocities and strain-rates in the Euro-Mediterranean region: from massive GPS data processing. Front Earth Sci 10:907897
14. Barzaghi R, Borghi A, Carrion D, Sona G (2007) Refining the estimate of the Italian quasi-geoid. Boll Geod E Sci Affin 66:145–159
15. Carminati E, Di Donato G (1999) Separating natural and anthropogenic vertical movements in fast subsiding areas: The Po plain (N.Italy) case. Geophys Res Lett 26:2291–2294
16. Alberti T, Anzidei M, Faranda D, Vecchio A, Favaro M, Papa A (2023) Dynamical diagnostic of extreme events in Venice lagoon and their mitigation with the MoSE. Sci Rep 13:10475
17. Faranda D, Ginesta M, Alberti T, Coppola E, Anzidei M (2023) Attributing Venice Acqua Alta events to a changing climate and evaluating the efficacy of MoSE adaptation strategy. NPJ Clim Atmospheric Sci 6:181
18. Yunus AP, Avtar R, Kraines S, Yamamuro M, Lindberg F, Grimmond CSB (2016) Uncertainties in Tidally adjusted estimates of sea level rise flooding (Bathtub model) for the Greater London. Remote Sens 8:366
19. Anderson TR, Fletcher CH, Barbee MM, Romine BM, Lemmo S, Delevaux JMS (2018) Modeling multiple sea level rise stresses reveals up to twice the land at risk compared to strictly passive flooding methods. Sci Rep 8:14484

Chapter 7
Natural Risk Assessment and Management in in Hexagonal France and Europe

Stephane Guillot and Tanguy Jonville

Abstract Hexagonal France's diverse geography, from mountains to coastlines, contributes to a high prevalence of natural risks, affecting most of the population. Climate change is exacerbating these risks, with rising temperatures leading to more frequent and intense events like heatwaves, storms, floods, and wildfires. In this paper, we detail specific risks in Hexagonal France. Coastal Risks: erosion, flooding, and storms threaten coastal areas, driven by rising sea levels and human activity. Mountain Risks: the Alps and Pyrenees experience heavy rainfall, increasing the risk of floods and landslides. Other Risks: wildfires are a major concern in southern France, while earthquakes and tsunamis pose threats, particularly in the Mediterranean region. France has a risk management framework involving prevention plans, local and national response coordination, and civil protection forces. However, the increasing frequency and severity of these events, coupled with budget constraints, necessitate a rethinking of risk prevention policies and better integration of scientific knowledge and socioeconomic factors. International cooperation, particularly within the Euro-Mediterranean region and the EU, is crucial for addressing these shared vulnerabilities.

7.1 Introduction

Overall, France's orography is highly diverse, ranging from the rugged peaks of the Alps and Pyrenees to the gentle rolling hills and plains of the central and northern regions including two maritime facades. This diversity and high-density population play a crucial role in the natural risks that France is facing. According to the Ministry

S. Guillot (✉)
University of Grenoble Alpes, University of Savoie Mont Blanc, CNRS, IRD, UGE, ISTerre, Grenoble, France
e-mail: stephane.guillot@cnrs.fr

T. Jonville
Sorbonne University, CNRS, LATMOS, Paris & Ecole Des Ponts, Marne La Vallée, France

© The Author(s), under exclusive license to Springer Nature Switzerland AG 2025

M. F. Buongiorno et al. (eds.), *The Challenge of Environmental Security in the Euro-Mediterranean Region*, NATO Science for Peace and Security Series C: Environmental Security, https://doi.org/10.1007/978-3-031-99646-7_7

for Ecological Transition, two-thirds of France's 36,000 communes are exposed to at least one natural hazard. Around six out of ten French people live in a commune exposed to a natural climate risk. Only 8000 communes in France, or 3.6 million inhabitants, are spared from natural risks. These are rural communities with almost four times fewer inhabitants and 2.5 times fewer urbanised areas than the national average. For some hazards, such as earthquakes, tsunamis and volcanism, the risks are low to moderate, but with climate change, the same cannot be said for other risks. Average temperatures in France have increased over the past century. France's climate has warmed by about 1.7 °C since the beginning of the twentieth century [1], which is higher than the global average. This is contributing to hotter summers, milder winters, and more frequent heatwaves. According to the last *IPCC report* [2], global warming is expected to reach 2.2 °C in France in 2030 compared with the pre-industrial era. This warming will continue until global carbon neutrality is achieved. The near future of the climate is therefore already almost written. After 2030, two main scenarios emerge, if governments meet the commitments set out in their national contributions to the Paris Agreement, global warming will be 2.8 °C by 2100 and ca. 3.5 °C in France. The announcements made since then by certain countries do very little to change this projection. If governments maintain their global policies in place at the end of 2025, global warming will reach 3.2 °C by 2100 and 4 °C in France. So far, this is the most likely scenario, given the current trend. Taking into account that since the end of the 2000s, there has been an average fourfold increase in serious events due to climate change in France compared with the period 1950–1999, we can expect an acceleration of major events in France over the next 30 years. To remain effective, risk prevention policies need to be rethought in the light of climate projections and adaptation issues. A better link is therefore needed between the production of knowledge on the one hand and the development and implementation of public policies on the other. Furthermore, risk prevention cannot be approached from an exclusively technical angle, and the Human and Social Sciences must be mobilized to take account of socio-economic perspectives and acceptability issues. Against a backdrop of budget restrictions and a sharp decline in public engineering, mobilizing the expertise of all knowledge-producing public establishments is becoming a major challenge. Given its position in the Euro-Mediterranean region and its diversity of orography, tectonics and microclimate, France's approach to assessing and managing natural risks has broader implications for neighboring countries that share similar vulnerabilities. In this paper, we present and discuss the orography of France and details the natural risks involved. Then we will discuss France's risk management doctrine and its organization. Finally, will broaden the debate to the Euro-Mediterranean basin.

7.2 France's Natural Risk Landscape

Hexagonal France, like many European countries, has a varied topography including two maritime facades, two active orogenic systems, large river basins and a dormant volcanic zone (Fig. 7.1). The French coastline stretches for around 5500 km, facing the Channel, the Atlantic Ocean and the Mediterranean Sea. The French coastline offers a remarkable diversity of landscapes and geological formations. The rocky coasts represent around 41% of the coastline, featuring cliffs, points, coves and bays. They are characteristic of plateau and result from the erosion of solid rock by waves and wind. The sandy coasts cover around 35% of the French coastline, they are made up of beaches, dunes and marshes. Finally, the salt and sea marshes represent 24% of the coastline, these areas are made up of mudflats and marshes, often used for growing salt. Natural coastal risks in France are a significant concern due to a combination of environmental, social, and economic factors. These risks are mainly driven by climate change and human activity (industry and tourism). Rising sea levels > 3 mm/year [3] with storm surges and coastal flooding caused by climate change and thermal expansion of water, threaten coastal regions by eroding beaches, flooding low-lying areas, and affecting infrastructure.

The projected rise in sea level for 2081–2100 relative to the period 1986–2005 varies between 26–55 cm and 52–98 cm^2. The northern Atlantic coast and the south-western Atlantic coast are heavily affected by erosion, resulting in the loss of beaches and damage to coastal infrastructure, particularly during winter. Coastal flooding is becoming more frequent especially in low-lying regions. Areas such as the Camargue and the Loire estuary are at risk, with marshes and floodplains being particularly vulnerable to sea-level rise and heavy rains. Storms affect a large part of coast. Winds gusting up to 200 km/h can occur anywhere in the country, causing many casualties and considerable damage. Winter storms particularly affect western France. On the coast, a storm combined with high tides can lead to flooding from the sea (marine submersions). This was the case in December 1999 (Lothar and Martin storms) and in February 2010 with Xynthia storm, which hitted Atlantic coasts. In 2020, 10% of France's population will live in coastal areas, whereas coastal municipalities will account for only 4% of the country's surface area. In coastal areas, more than 1400 municipalities have been identified as being at risk of flooding from marine submersion, affecting around 2% of the French population [3]. The ever-increasing pressure of tourism in coastal areas increases the vulnerability of these areas. Soil sealing also amplifies the phenomenon, by preventing water from infiltrating properly. At the scale of the Hexagon Hedou et al. [6] estimate that before 2050, ca. 4000 homes and 1000 business premises could be affected by coastal erosion, representing a total value of ca. 1 Md€.

With 25% of its surface area, France is a mountainous country. The Alps form one of the most prominent mountain ranges in Europe, and the southern part of this range stretches across southeastern France. The still tectonically active Alps form the natural border with Switzerland and Italy. The climate of the Alps is influenced by the Atlantic system from the west and the Mediterranean system from the Gulf of Genoa which brings abundant rainfall. Despite its generally sunny and nice climate on the Mediterranean coast, the municipalities in this region are particularly exposed to heavy rains and frequently affected by floods [7, 8], resulting in significant human and material losses. The Mediterranean Sea is an important water reservoir from which wet airmasses rise by meeting the hot air above the coastal cities, favoring intense convection phenomena, leading to short but heavy and localized rainfall. Wet

Fig. 7.1 Topographic map of France and surrounding countries including high-risk areas (modified from [10])

air masses coming from the sea can also be driven upwards confronted with the local relief, leading to stationary and sometimes heavy orographic rainfall [7, 8]. During the October 2015 flood, stationary and highly intense rainfall affected the coastal part of southeastern France, with 160 mm recorded in less than 2 h in Cannes, causing flooding of several rivers and associated landslides, resulting in catastrophic human and material losses (20 victims and 600 M€ of damage insured). The Alex storm in

October 2020 brought 80–100 mm of rain in 1 h, with total rainfall exceeding 600 mm in 24 h in the Vésubie and Roya valleys, causing flooding as well as mudslides and landslides [7, 8], 18 deaths and around 1 Md€ in material damage [9].

The Pyrenees form the natural border between France and Spain, extending from the Atlantic Ocean in the west to the Mediterranean Sea in the east (Fig. 7.1), and they are still tectonically active. The range has rugged peaks, with Aneto being the highest at 3404 m.

It is also known for its deep gorges and high mountain passes. The climate is dominated by the Atlantic influence from the west, which brings abundant rainfall, while the eastern part is becoming increasingly arid [11]. A systematic dataset of flood episodes referred to the entire Pyrenees massif which covers the period 1981–2015 show 181 flood events and 154 fatalities [12].

The Massif Central is an ancient mountain range and highland plateau located in the southern part of the country (Fig. 7.1). The region is known for its many dormant volcanoes, valleys, and deep gorges especially on this southeastern flank. It is made up of several massifs, such as the Monts Dore, the Cantal. The Cévennes massif massif is particularly exposed to intense rainfall (Cévenol episodes) caused by humid air from the Mediterranean [13].

Corsica's mountainous geography is characterized by rugged terrain, deep valleys, and high peaks. The island's central region is dominated by the Corsican Range. At lower elevations (below 1000 m), the climate is Mediterranean, characterized by warm, dry summers and mild, wet winters. Into the mountains (above 1500 m), the climate becomes more continental, with cooler temperatures and increased precipitation, especially in winter. During the night of 17–18 August 2022, a line of thundery squalls formed near the Balearic Islands, moving rapidly north-eastwards and affecting Corsica, Italy, Slovenia, Austria and then the Czech Republic in just 12 h. Wind gusts of up to 225 km/h in Corsica killed 12 people and injured 106 others, as well as causing extensive material damage. This event falls into the "derecho" category: a complex of thunderstorms extending over more than 100 km and characterised by the production of violent wind gusts in excess of 90 km/h over a wide area [14].

In between the mountain massifs, France orography is characterized by the Paris basin and the Aquitaine basin covering 32% of the surface area of the hexagon (Fig. 7.1) and two major valleys: the Rhine valley and the Saône-Rhône valley. This low-land area including Brittany and Normandy experience regular riverine flooding. The increasing intensity and frequency of these events are attributed to climate change but also urbanization. In terms of the number of French municipalities affected and the economic cost of disasters, flooding is the main risk to the country (Fig. 7.2).

Some 23,000 municipalities have been declared at risk of flooding by the prefects, covering around 64% of the country. Municipalities are confronted with flooding from overflowing rivers on the banks of major rivers (Garonne, Loire, Rhine, Rhône-Saône, Seine). Around a quarter of the French population is at risk of flooding from rivers overflowing their banks. In coastal areas, more than 1400 communes have been identified as being at risk of flooding from marine submersion, affecting around 2% of the French population [15]. In January 2024, heavy rainfall caused rivers to overflow

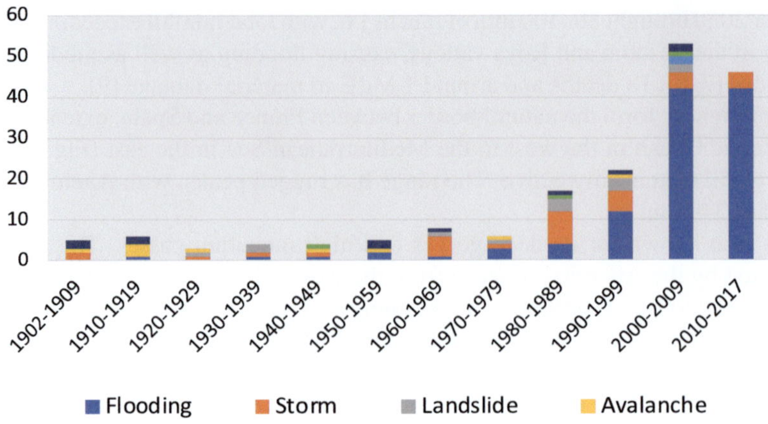

Fig. 7.2 Damaging natural events record in France (modified from the Ministry of Ecology)

and flood areas of northern France, affecting around 1300 homes, while almost 400 people have been evacuated. Several municipalities in the area have been impacted by power outages and water restrictions.

7.3 Other Risks in France

7.3.1 Wildfire

Southern France is highly vulnerable to wildfires due to its Mediterranean climate. Hot, dry summers combined with human activities and natural vegetation create conditions conducive to fire outbreaks. The number of fires and the surface burnt by them have significantly increased in the second half of the twentieth century [16], as a result of a drastic land abandonment and rural depopulation, leading to vegetation regrowth and fuel build-up [17]. The wildfires in France were also part of a broader trend of extreme weather events linked to climate change. In 2022 and 2023, France experienced devastating wildfires, particularly in the summer months, exacerbated by extreme heat and drought conditions. The most significant fires occurred in south-western France, and other areas in the south, such as Corsica. These fires led to the evacuation of thousands of residents and tourists. The Gironde wildfires, starting in mid-July 2023, spread rapidly and destroyed over 70,000 hectares of forest. Thousands of firefighters worked alongside aerial units to contain the blazes. The fires also led to major disruptions, including road closures and travel delays, affecting both locals and the summer tourist season.

7.3.2 Earthquake and Tsunami Risks

Apart from the Paris and Aquitaine Basins, the whole country is affected by moderate, diffuse seismicity (Fig. 7.3). The most active areas are the Pyrenees and Alps mountains. A lower seismic activity extends to northeastern France (Jura, Alsace, Vosges). Sparser seismicity is recorded in central and western France, in the Armorican Massif and the Massif Central. The seismicity in France is characterized by ca. 10 earthquakes of magnitude $Mw \geq 3.0$ every year, and slightly less than one $Mw \geq 4.0$ per year. Medium to large ($Mw > 5$) instrumental earthquakes have been recorded in all regions with background seismicity [18]. The most notable include: the Arette earthquake (Pyrenees, 1967, $Mw = 5.2$), Corrençon-en-Vercors (Alps, 1962, $Mw = 5.5$), along the Ligurian margin (1963, $Mw = 6.0$), Oléron (western France, 1972, $Mw = 5.0$). Most recently, the November 11, 2019, Le Teil earthquake (near Montélimar), $Mw = 4.9$, serves as a reminder that significant seismicity may also occur along the Rhone Valley [19]. The Fig. 7.3 also show that the southern part of Europe, specifically Italy, Greece and Turkey is highly exposed to Earthquake. The Santorini seismic crisis in winter 2025 underscores the importance of monitoring this activity.

The whole Mediterranean coasts, from Spain to Turkey have suffered from tsunamis in the past [22] (Fig. 7.3). In this geographical location, three main regions

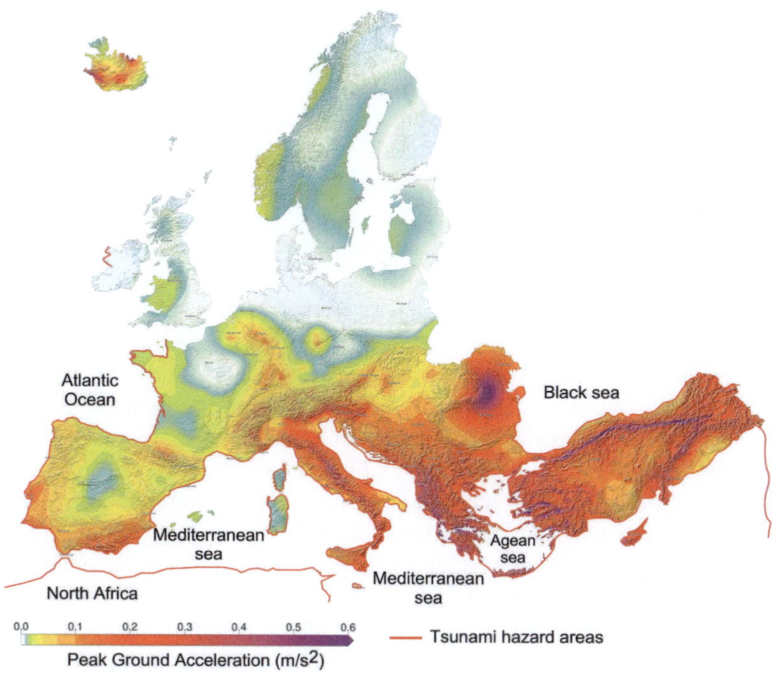

Fig. 7.3 Earthquake hazard map showing the potential ground shaking due to future earthquakes in Europe [20]. The red line along the costs corresponds to the European-wide tsunami risk [21]

are identifiable sources for tsunamis: (1) the region of the Ligurian sea (Italy) where seismic activities have triggered many devastating tsunamis, such as the earthquake-induced tsunami of 1887; (2) Sicily and the Aeolian Islands (Italy) which were the sources of numerous tsunamis triggered by- eruption-induced submarine landslides [23], or triggered by earthquakes, as was the case of Catania in 1693 [24]; and (3) the North-African margin which was a source that triggered several noticeable tsunamis, such as in 1365 and in 1856 [25]. More recently, in 2003, an earthquake of Mw = 6.9 occurred in the region of Boumerd'es-Zemmouri, Algeria, generating a tsunami which reached the Western Mediterranean coasts in a time span of a few minutes to several hours [26]. The Atlantic cost was historically hit by Tsunami in 1785, 1875 and 1882 tsunamis in Charente Maritime (Fig. 7.3). Comparison of the documentary sources showed that the origin of these tsunamis did not coincide with local earthquakes. In each case, the noted effects correspond to relatively abrupt rises and falls of the sea without it being possible to assign them to a seismic origin, not even a remote one. The potential sources of these small Tsunami are probably earthquake of Mw > 6.0 generated along the Mid-Oceanic Atlantic Ridge as the French Tsunami Warning center (CENALT) recorded since 2012.

7.4 Risk Assessment Framework in France

In France, risk prevention is articulated with land-use planning policies through the Risk Prevention Plans (Plans de Prévention des Risques, PPR) issued by the State. It implements regulations based on hazard mapping, ranging from specific construction standards to outright building prohibitions. The initial response to a crisis is generally local and relies on the mayor and the departmental prefect (Fig. 7.4).

The mayor is responsible for directing operations within the Communal Command Post (PCC), until the departmental prefect intervenes. Even then, the mayor is responsible to his constituents for safeguard measures. For this reason, municipalities under a PPR must set up a PCS (Plan Communal de Sauvegarde). The aim of this plan is to prepare the response to a crisis. It organises the dissemination of warnings and safety instructions, as well as safeguard and protection measures, and lists the resources available. It thus ensures the continuity of the municipality's compulsory tasks. When the crisis affects more than one municipality, or when municipal resources are overwhelmed, the prefectoral services take over. First and foremost, it is the departmental prefect who is mobilized to organize and coordinate the emergency response. To do this, he relies on the SIDPC (Interdepartmental Civil Defense and Protection Service), which organizes the setting up of a Departmental Operations Centre (COD) and an Operational Command Post (PCO). When departmental resources are exceeded or when more than one department is affected, it is the Prefect of the Defense Zone who takes over, supported by the Zone Headquarters, and who is responsible for liaising with the national level. The Defense and Security Zone is a specific administrative area for civil defense and security. The departmental prefect and the zone prefect rely on ORSEC (Organisation de la Réponse de Sécurité Civile) plans, drawn up by

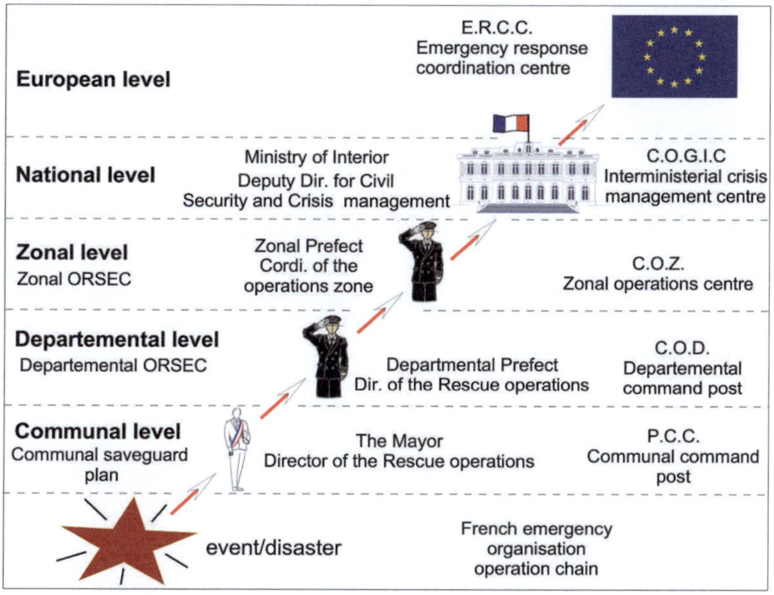

Fig. 7.4 Crisis management in France, from the communal level to European level (modified from [27])

their departments, to organise their response. Adapted to each type of anticipated crisis, the ORSEC plans organise the operational system, describe the anticipated risks and scenarios, and provide for the organisation of training and exercises to enable all those involved to develop operational know-how. In the event of a major disaster, the State organises itself into a COGIC (Centre Opérationnel de Gestion Interministérielle de Crise—Operational Centre for Interministerial Crisis Management) to coordinate the response at national level in conjunction with the prefectoral levels. The Prime Minister is responsible for national defense under article 21 of the Constitution. Together with the government, he is accountable to parliament for the conduct of defense policy. The Prime Minister delegates operational management to the Minister of the Interior when the crisis occurs on national territory or to the Minister of Foreign Affairs for external crises. The minister responsible for operational management relies on all the ministries represented in the inter-ministerial crisis unit on behalf of the Prime Minister.

7.5 Civil Protections

Civil protection includes all civilian and military fire brigades (Fig. 7.5). There are around 220,000 civilian fire-fighters in France (professional and volunteer fire-fighters) under the aegis of the departmental fire and rescue services (SDIS). In

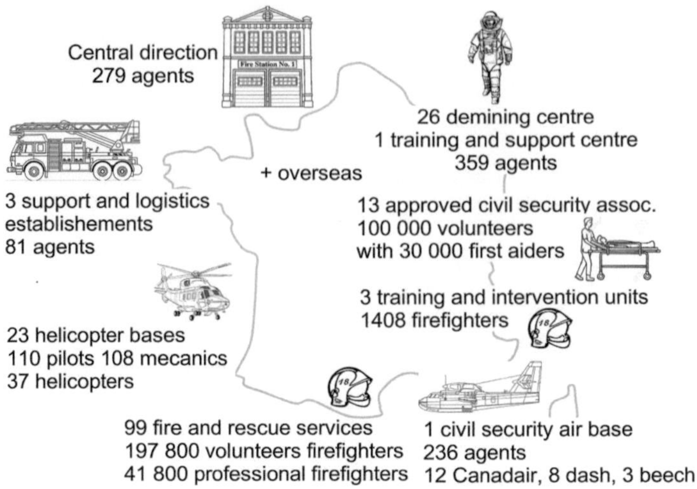

Fig. 7.5 Civil protection organisation in France (courtesy Bruno Ulliac, Direction Générale de la Sécurité Civile et Gestion des Crises)

addition, the Brigade des sapeurs-pompiers de Paris (French army) is at the disposal of the Prefect of Police of Paris with 12,000 agents and the Bataillon des marins-pompiers de Marseille (French Navy) is at the disposal of the Prefect of Police of Bouches-du-Rhône with 2400 agents. National military reinforcements, from the army, make up the military formations of the Civil Security (Unités d'instruction et d'intervention de la Sécurité civile, UIISC). The resources consist of fire and rescue service equipment, as well as Civil Security UIISC resources, Dragon helicopters and water bomber aircraft.

National Gendarmerie helicopters are also used for civil protection missions. The approved civil protection associations should not be forgotten in this overview (200,000 members, including a core group of 70,000 active first-aiders). The French Red Cross, Protection Civile, the Fédération Française de Secours et de Secourisme and many others, in addition to their day-to-day activities, provide considerable support during a crisis.

7.6 The Armed Forces

The armed forces are one of the players contributing to the State's response in crisis situations, primarily abroad, but also at home, whether in terms of military defense (territorial defense) or civil defense (public safety and civil security). The operational chain, commanded by the Armed Forces Chief of Staff, who reports directly to the President of the Republic, is structured according to a three-level hierarchy (strategic—operational—tactical). In mainland France, it is organized around the defense and security zone headquarters, the operational level of command, whose general officers are the contacts and military advisers of the zone prefects. In accordance with IM 10100 [27], the armed forces may be deployed to support or back up the internal security forces if requested to do so by the prefect and approved by the armed forces headquarters.

7.7 Academics and Researchers

According to the International Disaster Data base [28], Hexagonal France records 14% of the serious event since 1900 in Europe, slightly ahead of Italy. In order to mitigate this challenge, the French government asked in 2022, the various research organisms to structure themselves around joint projects. The aim of developing, over the next 5–10 years, knowledge and innovative tools to better detect, understand, quantify, anticipate and manage risks and disasters. However, researchers are only involved in crisis management on an ad hoc natural crisis. They are sometimes called upon by the Prefecture as experts on highly specialized subjects (the Wedenheim seismic crisis in Alsace between 2020 and 2021, the Teil earthquake in the Ardèche in 2019, monitoring major landslides—Séchilienne in Isère or La Clapière in the Alpes-Maritimes). While the universities and the Centre National de la Recherche Scientifique (CNRS) are attached to the Ministry of Research, the other research bodies are attached to the Ministry of Ecology. The Direction Générale de la Prevention des Risques (DGPR) relies on some 10 research bodies to implement the recommendations on risk mitigation and anticipation. There are many examples of public research bodies that are operational and called upon in the event of a specific crisis (Table 7.1), "Météo-France" is an important player, it is a meteorology research center associated with the CNRS and provides 7-day forecasting, participation in 50-year climate scenarios, which also plays a decisive role in weather warnings (wind, rain, snow, hail). Météo-France provides real-time weather forecasts and alerts, enabling timely responses to imminent threats such as storms (Fig. 7.6). It collaborates with the SHAPI (Service Central Hydrométéorologique d'Appui à la Prévention des Inondations), a department of the ministry of ecology responsible for flood forecast.

7.8 The Issue of Crisis Management Costs

As part of the risk prevention plan, the "Fonds de Prévention *des Risques Naturels Majeurs*" (FPRNM) or "Barnier" fund, set up in 1995, supports measures to prevent or protect people and property exposed to natural risks. It covers a rate of reimbursement for damage suffered by local authorities and/or private individuals and can be supplemented by funds mobilized by local authorities, as well as European funding.

Table 7.1 Principal establishments that manage natural risks

Establishment	Supervision	Risk assessment
BRGM, *Bureau des Ressources Géologiques et Minières*	Ministries of Ecology/ Research	Gravity risk, landslides, groundwater, seismic risk, coastal risk
CEREMA, *Centre d'études et d'expertise sur les risques, l'environnement, la mobilité et l'aménagement*	Ministry of Ecology	Infrastructure, hydraulic structures, coastal risks, methodological developments, network management
CSTB, *Centre Scientifique et Technique du Bâtiment*	Ministry of Ecology	Buildings and infrastructure
IGN, *Institut national de l'information Géographique et forestière*	Ministries of Ecology/ Research	Cartography
INERIS, *Institut national de l'Environnement Industriel et des Risques*	Ministries of Ecology/ Research	Industrial risk
INRAE, *Institut national de recherche pour l'agriculture, l'alimentation et l'environnement*	Ministries of Agriculture/ Research	Hydrology
Météo-France	Ministries of Ecology/ Research	Climate change, precipitation, weather forecast
SHOM, *Service Hydrographique et Océanographique de la Marine*	Ministry of the Army	Hydrography, oceanography
ONF, *Office National des Forêt*	Ministry of Agriculture	Mountain risks, wildfire
IPGP, *Institut de Physique du Globe de Paris*	Ministry of Research	volcanology
CEA, *Commissariat à l'Energie Atomique*	Ministries of Research/Army/ Energy/industry	Tsunami

The FPRNM is funded by a levy on all insurance contracts ('natural disasters' regulatory contribution). With a budget of 225 M€ in 2025, the FPRNM is well short of what is needed. In 2021, "France Assureurs" estimates that 3.05Md€ will be paid out in compensation for natural disasters. 'The increasing value of real estate and the growing number of claims are prompting insurers to withdraw from certain areas, which is having an impact on the sustainability of natural catastrophe insurance. The "Caisse Centrale de Réassurance" predicts that the costs associated with all hazards will increase by 50–85% between now and 2050, depending on the climate scenario adopted. In the case of forest fires, the 2.5 Md€ Green Fund for 2024 is used to finance prevention measures aimed at the energy transition. With regard to crisis management assistance, recent crises over the last 10 years, and in particular the fires in the summer of 2022, have revealed the need for a new civil protection strategy. These

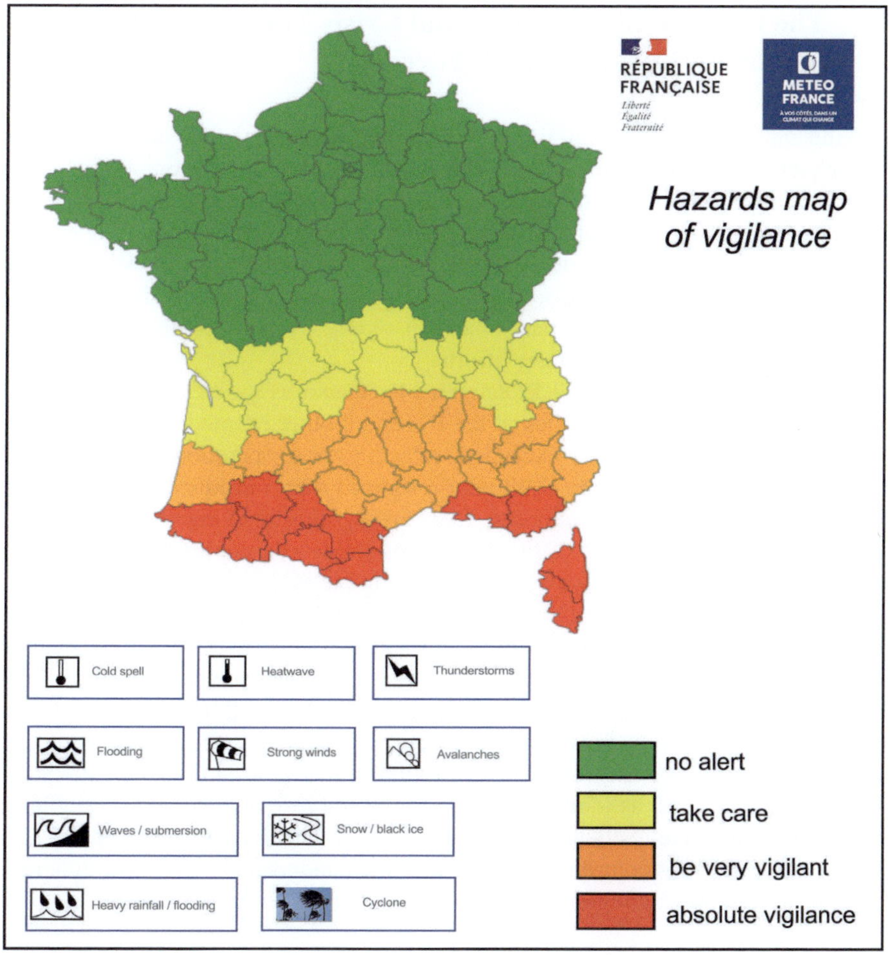

Fig. 7.6 Hazards map of vigilance launched by Météo-France (television, social networks, website) showing the degree of vigilance from green (no alert) to red (absolute vigilance). Pending of warning type (cold spell, flooding, heatwave ...) a specific map is proposed. (Modified from [29])

investments, scheduled over 5 years, involve new equipment with the renewal of the helicopter fleet, the reinforcement of material and human resources, with the launch of a plan to reinforce and modernize the capacities of the Civil Security Military Training.

7.9 The Concept of Supranational Crisis Management Solidarity/European Task Force

Satellite data are now essential for anticipating and managing natural hazards throughout Europe. Copernicus is the Earth Observation component of the European Union's Space Program, which monitors our planet and its environment and provides with actionable information from satellite and in situ data. The Copernicus Emergency Management Service (CEMS) supports disaster response by providing geospatial information and maps. It issues early warnings for events like floods and wildfires and can be activated on demand to supply maps to facilitate evacuation planning and post-disaster recovery efforts.

The European Union's Civil Protection Mechanism (Fig. 7.7) offers enhanced cooperation between the countries of the EU and 9 other Member States when the resources of one of them are exceeded, affecting their operational response and enabling them to benefit from additional resources (first aid kits, accommodation equipment, water pumps, generators, fuel, fire-fighting resources, etc.). The EU contributes 75% of the costs of these missions (transport, deployment and operational costs). Following a request for assistance in the event of a major disaster, this mechanism enables the activation of the Emergency Response Coordination Centre (ERCC), which monitors events around the world 24 h a day. In 2022, the mechanism was activated 106 times. France made little use of the mechanism, with storm Irma and the mega forest fires in Gironde in 2022. France's robust approach to natural risk assessment provides valuable insights and best practices for the Euro-Mediterranean region, which shares many of its vulnerabilities. France's participation in cross-border initiatives, such as joint training exercises and resource sharing, strengthens the region's collective resilience to natural disasters. The dissemination of research findings and risk maps contributes to a more unified understanding of hazards across the Euro-Mediterranean region.

7.9.1 Conclusion

Natural risk assessment in France stands as a dynamic and integral element of the nation's approach to disaster prevention and resilience building. It combines rigorous scientific methodologies, comprehensive data collection, and multi-level governance to identify, evaluate, and mitigate a wide range of hazards—from floods and landslides to seismic events and extreme weather. This systematic process not only informs emergency preparedness and public safety strategies but also drives continuous improvement in response planning and community engagement. As climate change and urban development introduce new challenges, France's commitment to advancing its risk assessment frameworks remains essential for protecting lives, sustaining the environment, and ensuring a resilient future. Solidarity and organization between the countries of Europe and the Mediterranean basin is essential if we

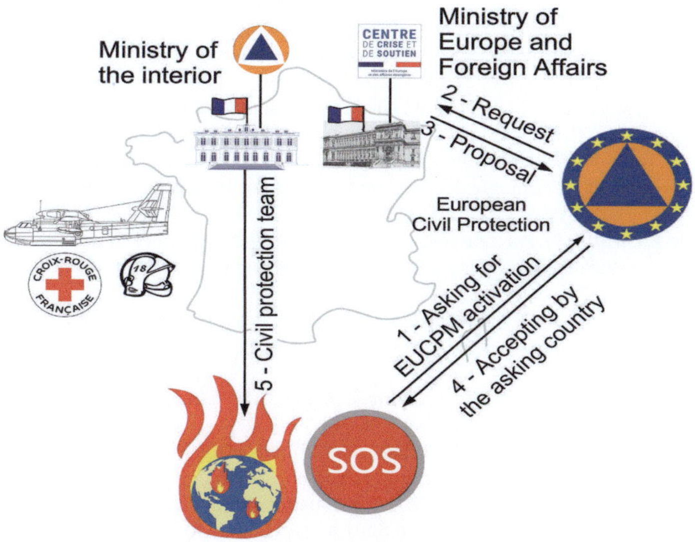

Fig. 7.7 Scheme of the European Protection mechanism. Whenever a European country or a country outside Europe requests help from Europe, European civil protection contacts all the member states via their foreign affairs ministers. The civil protection authority of the country requested makes a proposal to European civil protection, which coordinates the responses of the Member States and forwards the proposal to the requesting country, which accepts or rejects it. If the country accepts the proposal, the coordinated resources are deployed to the crisis zone

are to cope with the growing risks associated with climate change. The contribution of Copernicus is essential to better anticipate and manage natural hazards while the European Union's Civil Protection Mechanism is still efficient to share our resources and expertise.

Acknowledgements This work is supported the CNRS-INSU and the Ministry of Ecology

References

1. Ribes A, Boé J, Qasmi S et al (2022) An updated assessment of past and future warming over France based on a regional observational constraint. Earth Syst Dyn 13:1397–1415
2. IPCC (2023) Climate change 2023: synthesis report. In: Contribution of working groups I, II and III to the sixth assessment report of the intergovernmental panel on climate change, (core writing team Lee H, Romero J). Geneva, Switzerland, pp 35–115
3. Cazenave A, Le Cozannet G (2014) Sea level rise and its coastal impacts. Earth's Futur. https://doi.org/10.1002/2013EF000188
4. IM10100. Instruction ministérielle n°10100 du 14/11/2017 relative à l'engagement des armées sur le territoire national.www.vie-publique.fr/eclairage/273322-climat-vers-une-aggravation-des-risques-naturels

5. Gaume E, Borga M, Llassat MC et al (2016) Mediterranean extreme floods and flash floods. In: The Mediterranean region under climate change. A scientific update. IRD Editions, pp 133–144. Coll. Synthèses 978-2-7099-2219-7
6. Hedou F, Herman A, De Luca H et al (2024) Projection du trait de côte et analyse des enjeux au niveau national Horizons 2050 et 2100, Rapport-Cerema_Etude-2050-2100
7. Chmiel M, Godano M, Piantini M et al (2022) Brief communication: Seismological analysis of flood dynamics and hydrologically triggered earthquake swarms associated with storm Alex. Nat Hazards Earth Syst Sci 22:1541–1558
8. Drobinski P, Da Silva N, Panthou G et al (2018) Scaling precipitation extremes with temperature in the Mediterranean: past climate assessment and projection in anthropogenic scenarios. Clim Dyn 51:1237–1257
9. IGEDD (2023) Retour d'expérience des intempéries des 2 et 3 octobre 2020 dans les Alpes-Maritimes—Enseignements de la crise et propositions pour une reconstruction résiliente
10. www.scilands.de/en-gb
11. Cuadrat JM, Serrano-Notivoli R, Prohom M et al (2024) Climate of the Pyrenees: extremes indices and long-term trends. Sci Total Env 933:173052
12. Llasat MC, Llasat-Botija M, Pardo E et al (2024) Floods in the Pyrenees: a global view through a regional database. Nat Hazards Earth Syst Sci. https://doi.org/10.5194/nhess-24-3423-2024
13. Ceresetti D, Ursu E, Carreau J et al (2012) Evaluation of classical spatial-analysis schemes of extreme rainfall. Nat Hazards Earth Syst Sci 12(11):3229–3240. https://doi.org/10.5194/nhess-12-3229-2012
14. Fery L (2024) Faranda D (2024) Analysing 23 years of warm-season derechos in France: a climatology and investigation of synoptic and environmental changes. Weather Clim Dynam 5:439–461. https://doi.org/10.5194/wcd-5-439-2024
15. Boccard N (2018) Natural disasters over France a 35 year assessment. Weather Clim Extrem 22:59–71
16. Moreno JM, Vazquez A, Velez R (1998) Recent history of forest fires in Spain. In: Moreno JM (ed) Large forest fires. Backhuys Publishers, Leiden, The Netherlands, pp 159–186
17. Le Houérou HN (1987) Vegetation wildfires in the Mediterranean basin: evolution and trends. Ecol Mediterra 13(4):13–24
18. Mazzotti S, Jomar H, Masson F (2019) Processes and deformation rates generating seismicity in metropolitan France and conterminous Western Europe. BSGF— Earth Sci Bull 191(19). https://doi.org/10.1051/bsgf/2020019
19. Ritz JF et al (2020) Surface rupture and shallow fault reactivation during the 2019 Mw 4.9 Le Teil earthquake, France. Commun Earth Environ 1(1):1–11
20. www.efehr.org
21. www. preventionweb.net/files/3831_ Tsunamihazard N3.jpg
22. Sahal A, Roger J, Allgeyer S et al (2009) The tsunami triggered by the 21 May 2003 Boumerd'es-Zemmouri (Algeria) earthquake: field investigations on the French Mediterranean coast and tsunami modelling. Nat Hazards Earth Syst Sci 9:1823–1834
23. Gerardi F, Barbano MS et al (2008) Discrimination of tsunami sources (earthquake versus landslide) on the basis of historical data in Eastern Sicily and Southern Calabria. B Seismol Soc Am 98(6):2795–2805
24. Gutscher MA, Roger J, Baptista MA et al (2006) Source of the 1693 Catania earthquake and tsunami (southern Italy): new evidence from tsunami modelling of a locked subduction fault plane. Geophys Res Lett 33 L08309. https://doi.org/10.1029/2005GL025442
25. Lasset PJ, Hébert H, Maouche S et al (2006) The tsunami induced by the 2003 Zemmouri earthquake (Mw=6.9, Algeria): modelling and results. Geophys J Int 166:213–226
26. www.meurthe-et-moselle.gouv.fr/Actions-de-l-Etat/Securite-et-protection-de-la-population/Securite-civile/Gestion-de-crises/Gestion-de-crises
27. www.igedd.developpement-durable.gouv.fr/retour-d-experience-des-intemperies-des-2-et-3-a35.html.
28. www.emdat.be/
29. www.data.gouv.fr/fr/reuses/alerte-meteo

Chapter 8
Harnessing Satellite and Digital Technologies for Climate Action: The Case of Jordan

Ruba Ajjour

Abstract This paper seeks to highlight the role of satellite and digital technologies in climate action, with a focus on the Jordanian context. It also outlines national vulnerabilities, recent advancements in climate modeling and data analysis, and presents applied case studies, including urban resilience planning and AI-driven climate forecasting. Moreover, it explores how the integration of satellite and digital technologies, including artificial intelligence (AI), can play a transformative role in monitoring, managing, and mitigating climate risks in Jordan. The paper concludes with recommendations for overcoming challenges and scaling technology integration in climate governance and adaptation strategies.

Keywords Climate change · Jordan · Satellite technologies · Digital transformation · AI · Adaptation · Resilience · NDC · Water scarcity · Climate risk

8.1 Introduction

Jordan is among the most climate-vulnerable countries in the Middle East and North Africa (MENA) region. Its water scarcity, rapid urbanization, and socio-political challenges intersect to exacerbate the climate crisis. Jordan's limited natural resources, particularly water, combined with demographic pressures and regional instability, necessitate robust climate action strategies.

Jordan's climate action is guided by a whole-of-government approach, with the Ministry of Environment serving as the national focal point for climate negotiations and implementation.

In recognition of these pressing challenges, Jordan has taken significant steps toward mainstreaming climate change into its national policy framework. The country

R. Ajjour (✉)
Climate Change Studies Division, Royal Scientific Society, Amman, Jordan
e-mail: ruba.ajjour@rss.jo

© The Author(s), under exclusive license to Springer Nature Switzerland AG 2025
M. F. Buongiorno et al. (eds.), *The Challenge of Environmental Security in the Euro-Mediterranean Region*, NATO Science for Peace and Security Series C: Environmental Security, https://doi.org/10.1007/978-3-031-99646-7_8

ratified the Paris Agreement and has since submitted two updated Nationally Determined Contributions (NDCs), with the latest (2021) committing to a 31% reduction in greenhouse gas emissions by 2030, contingent on international support. Additionally, Jordan has formulated its National Climate Change Policy 2022–2050, which articulates a vision for a climate-resilient and low-emission future. In terms of Adaptation, Jordan developed the National Adaptation Plan (NAP), in alignment with the UNFCCC framework, prioritizes adaptation actions across water, agriculture, energy, health, and infrastructure sectors. Furthermore, Jordan has initiated cross-sectoral coordination mechanisms involving line ministries, municipalities, academia, and civil society to ensure an inclusive and integrated climate response.

Jordan also actively engages in international cooperation and regional platforms to mobilize technical and financial support for climate resilience. Strategic partnerships with UN agencies, donor countries, and international research institutions have enhanced Jordan's capacity to conduct vulnerability assessments, pilot adaptation projects, and access climate finance. The integration of digital technologies, satellite data, and AI tools into its climate governance framework is a testament to the country's commitment to leveraging innovation for sustainable development.

8.2 National and Sectoral Climate Context

Jordan's geographic and climatic characteristics expose it to significant environmental stress. Over 75% of its land area is classified as arid or semi-arid, receiving less than 200 mm of rain annually. With one of the lowest levels of renewable water resources per capita globally, the country ranks among the world's most water-scarce nations.

According to Jordan's Fourth National Communication to the UNFCCC (2023) report, a 2.1 °C rise in temperature and a 21% decline in annual precipitation, with increased extreme weather events. This impacts water availability, food security, biodiversity, public health, and infrastructure. The agriculture sector, which consumes over half the country's water resources, is particularly affected. Climate change has intensified these vulnerabilities, leading to more frequent droughts, rising temperatures, and altered precipitation patterns.

These trends are accompanied by increased frequency and intensity of extreme weather events such as heatwaves, flash floods, and prolonged droughts. These climatic changes have direct and far-reaching impacts on various sectors, notably water availability, food production, biodiversity, public health, energy, and infrastructure resilience.

The agriculture sector is particularly vulnerable, consuming over 52% of the nation's water resources. Changing rainfall patterns and higher evapotranspiration rates threaten crop yields and livestock production, endangering food security and rural livelihoods. Urban areas, which host around 90% of Jordan's population, face additional challenges such as water supply interruptions, increased heat stress, and the overstretching of basic services.

Climate change has intensified existing environmental vulnerabilities, accelerating land degradation, desertification, and the loss of ecosystem services. Biodiversity hotspots, particularly in the Jordan Rift Valley and mountainous areas, are at risk of habitat loss. Health systems are increasingly burdened by the spread of climate-sensitive diseases, heat-related illnesses, and air pollution.

In response, Jordan is expanding its efforts in climate risk assessment, using advanced modeling and historical data analysis to map vulnerabilities and inform adaptation strategies. This includes the development of climate indices and early warning systems that allow for more targeted and timely interventions across affected sectors.

8.3 Satellite, Digital Technologies and AI in Climate Action—Examples from Jordan

In the education and research domain, Jordan's research institutions and universities, including RSS, NARC, the University of Jordan and Jordan University of Science and Technology (JUST), have developed remote sensing and climate change programs and curricula. They engage in applied research projects involving drone imagery, land-use classification, and data visualization—helping to build the next generation of climate scientists and practitioners.

Hereafters are few noteworthy examples that represent diverse case studies to illustrate how Jordan is leveraging technology across different levels—national, municipal, and community—to create more adaptive and informed climate policies. They highlight the importance of local innovation, strategic partnerships, and the integration of climate data into both long-term planning and day-to-day operations.

The Jordan Meteorological Department (JMD), in collaboration with the European Centre for Medium-Range Weather Forecasts (ECMWF), has expanded the country's climate monitoring network by installing automated weather stations that feed into regional early warning systems. These systems are vital in issuing alerts on heatwaves, sandstorms, and flash floods, supporting public safety and preparedness.

Another noteworthy example is the development of climate-resilient urban master plans for the municipalities of Irbid and Mafraq. Supported by UN Habitat and implemented by the Royal Scientific Society (RSS) and LDK this initiative integrates climate data with spatial planning to enhance urban resilience. The plans include climate adaptation strategies, disaster risk reduction measures, and gender-sensitive approaches.

Another flagship project is "Jahez" (Arabic word translates to Ready in English), launched at COP29 as part of the Climate-Refugee Nexus Initiative introduced by King Abdullah II. Jahez aims to strengthen climate resilience in refugee-hosting communities through enhanced forecasting systems, nature-based solutions, and integrated early warning mechanisms. Partners include the UK Met Office, IWMI, and various Jordanian ministries and municipalities.

Moreover, AI technologies are increasingly being explored in Jordan to support climate action. These include predictive analytics for early warning systems, machine learning algorithms for climate data analysis, and tools for greenhouse gas (GHG) inventory estimation, particularly Scope 3 and Scope 4 emissions.

8.4 Challenges and Enabling Environment

Jordan's National Strategy for AI (2023–2027) outlines the need for an enabling ecosystem to support AI adoption across sectors. While promising, AI deployment is still constrained by limited institutional capacity, fragmented data systems, and funding challenges.

Despite progress, several challenges hinder the full integration of satellite and digital technologies in climate policy and practice. These include:

- Technical and institutional capacity gaps within public institutions
- Reliance on external expertise for complex modeling and data interpretation
- Limited infrastructure for data integration and storage
- Inadequate funding for research and implementation

Creating an enabling environment requires strengthening local capacities, fostering innovation ecosystems, and establishing robust regulatory and governance frameworks. Public–private partnerships and international cooperation are key to bridging the digital divide and ensuring equitable access to climate technologies.

8.5 Recommendations and Future Directions

To accelerate the integration of digital technologies in climate action, the following recommendations are proposed:

- Develop a national digital climate platform to centralize and share climate data
- Support AI startups and innovation hubs focused on environmental solutions
- Invest in training and capacity building for local authorities and technical staff
- Integrate digital tools into national adaptation plans (NAPs) and nationally determined contributions (NDCs)
- Encourage cross-sector collaboration between government, academia, civil society, and the private sector

Jordan can also serve as a model for other MENA countries by piloting scalable solutions and contributing to regional knowledge-sharing initiatives.

8.6 Conclusion

Jordan's efforts to integrate satellite, digital, and AI technologies into its climate action agenda, highlight the transformative potential of innovation in building resilience. While challenges remain, targeted investments, partnerships, and policy coherence can support a sustainable digital transition. Leveraging these technologies will not only help Jordan meet its NDC targets but also enhance its role as a regional leader in climate-smart development.

References

1. Ministry of Environment (2021) Jordan's updated nationally determined contributions (NDC)
2. RICCAR, Mashreq Regional Climate Modeling Results.
3. UNFCCC, National communications and adaptation reports for Jordan
4. IPCC AR6 (2023) Climate change 2023: impacts, adaptation and vulnerability
5. Jahez Project Brief (2024) Ministry of planning and international cooperation
6. National AI Strategy (2023–2027) Ministry of digital economy and entrepreneurship
7. https://unhabitat.org/af-lebanon-jordan

Chapter 9
EO Integrated Services to Monitor and Mitigate Geohazards: Contribute from INGV Center for Space Observations of Earth

Maria Fabrizia Buongiorno, Giorgiana De Franceschi, Vincenzo Romano, Carlo Marcocci, Simone Atzori, Stefano Corradini, Simona Scollo, and Giuseppe Puglisi

Abstract The increasing population and human activities produce an impact on the terrestrial environment and increase the risk due to geohazards on infrastructures affecting the sustainability of development models for human society. As we live in a very global environment it is necessary to consider that extreme weather events, earthquakes, volcanic eruptions may produce an amplification of the effects both in space (territory) and in time (long term effects) dimensions. In Europe since the creation of the Coperinicus Monitoring System, former Global Monitoring Environment and Security (GMES). The Istituto Nazionale di Geofisica e Vulcanologia (INGV) has actively contributed in the last 30 years to the development of space technologies both in the Earth Observation sector (including all available Optical and SAR imaging systems) and in the Global Navigation Satellite Systems (GNSS) sector, allowing the INGV to play a role of excellence both for the study of satellite missions and for the development and supply of operational products for end-users. In 2020 INGV established the Center for Space Observations of Earth (COS) with the aim of coordinating the INGV activities in the Space and Aerospace sector. The COS contributes to the strategy of INGV Departments, namely Environment, Earthquakes and Volcanoes, and the INGV Sections/Volcano Observatories by adding specific products and services based on space observations to assist monitoring, surveillance, research and services for the society regarding Earth processes. In 2022 the National Recovery and Resilience Plan (PNRR) offered specific funding for the Space Sector which will permit to build, in the next 3 years, both the COS infrastructures (ICT platform) and new research activities to further develop the COS modules, as well as reinforce the cooperation with industrial partners by to provide a series of complementary activities which could improve the national expertise and technological transfer in

M. F. Buongiorno (✉) · G. De Franceschi · V. Romano · C. Marcocci · S. Atzori · S. Corradini · S. Scollo · G. Puglisi
National Institute of Geophysics and Volcanology (INGV), Rome, Italy
e-mail: fabrizia.buongiorno@ingv.it

© The Author(s), under exclusive license to Springer Nature Switzerland AG 2025
M. F. Buongiorno et al. (eds.), *The Challenge of Environmental Security in the Euro-Mediterranean Region*, NATO Science for Peace and Security Series C: Environmental Security, https://doi.org/10.1007/978-3-031-99646-7_9

with a consequent advantage for all the space sector chain supporting the mitigation and monitor of Geohazards at National and Global level.

Keywords Geohazards · Earth observation · Space weather · Geophysics · ICT platform · Technological transfer · Outreach

9.1 Introduction

Disaster risk is increasing across the globe and more people, and their assets, are increasingly vulnerable and exposed to that risk, especially in developing countries, where disasters worsen inequality and hinder progress toward sustainable development goals. This elevated risk, which is a sign of a clear resilience gap, is leading to larger and more frequent disasters with greater economic losses that are costing lives and livelihoods. Unsurprisingly, weather-related disasters were top of the most likely risks over the next decade, with 66% of respondents in the report ranking extreme weather as the greatest concern. The EU's Copernicus Earth observation program reported that 2023 was the hottest year on record, showing the increasing of extreme events occurrence such as flooding and wildfires, and warming Arctic temperatures [1, 2]. But non-weather-related events have also a very significant impact even if they rank lower down in the last decade due to the very rapid increase of extreme weather events. Earthquakes and volcanic eruptions, while not directly related to climate change, can cause serious disruption and devastation. More than 3700 people were made homeless in Iceland due to volcanic activity, the New York Times reported in January 2024, due to volcanic activity. In 2024, global seismic activity was lower than average, with 99 earthquakes exceeding magnitude 6, the fewest since 1982. There were also no earthquakes exceeding magnitude 8 for the third consecutive year. Meanwhile, 74 confirmed eruptions occurred from 65 different volcanoes, with 32 being new eruptions (Global Volcanism Program report 2024). Globally, natural disasters affected 167.2 million people, caused 16,753 fatalities, and resulted in US$241.95 billion in economic losses [3]. A geohazard is a type of geological condition that poses possibilities to cause significant damage to over the surface structures. These geological conditions are formed over the years through the effect of environmental changes along with variations in lithological temperatures and pressures. Consequently, a combination of various technologies is required to assess, identify, and combat the geological problems before the setup of new infrastructures to mitigate any risks. The disaster management services utilize different data collection methods to correctly assess technical and social issues and reduce the impact of a wide range of geological hazards such as volcanic activities, tsunamis, earthquakes, landslides, and floods.

Since the late 90 s, INGV effectively contributes to the development of space technologies both in the Earth Observation sector (including all available space systems Optical, SAR etc.) and in the GNSS sector allowing the INGV to play a role of excellence both for the study of satellite missions and for the development and supply of operational products for end-users. Since 2003 INGV was involved in the development of Copernicus Monitoring System by and active participation on National Steering Committee lead by the Ministry of Research and by contributing to the definition of the GMES Emergency Service for Earthquake and Volcanoes (2004–2012). The Centre for Earth Space Observations (COS) has been established in 2020 with the aim of linking and planning the activities of the Institute in the fields of space and aerospace, as well as to encourage the participation of researchers in designing and implementing services and products for scientific research and applications to support INGV research, external Public and Private Entities and society. In this area, INGV plays a particularly suitable role for cooperation with industrial and institutional partners by providing a series of complementary activities. In particular COS represents INGV in national working groups and committees operating in the Space and Aerospace sector, particularly those represented in the COMINT MUR, ASI, and internationally by ESA, EU, ICAO, COSPAR, UN, and other Space Agencies (NASA, CNSA, JAXA, CONAE, CSA, etc.). COS also represents INGV at the Copernicus National User Forum by supporting the identification of user requirements to develop the next generation of products and services based on EO data.

In 2021 the implementation of Italian PNRR (Recovery and Resilience Plan) after the COVID 19 crisis, offered the opportunity to support the operational development INGV COS [4] in the framework of the PNRR MEET project coordinated by INGV, where the ICT infrastructure called Platform of Earth Observation from Space (PEOS) to support COS was funded. PEOS infrastructure supports the development and operation of integrated processing chains to assimilate ground measurements, geophysical models and EO and other Space data to improve INGV monitor capabilities geo-hazards and Space Weather. INGV COS also participates in the SPACE-IT-UP project coordinated by the Polytechnic of Turin dedicated to the development of research and technical developments for future space systems and services including the training of students and young researchers in Space Science and Technology.

9.2 COS Management Structure a Proposed Development Plan

The COS was established in July 2020 by the INGV Board of Directors. Each INGV Centre has a Statute that describes its functions and organizational structure. The COS has a modular structure, which can be modified based on the strategic lines of the Organization and the development needs of products and services based on satellite data and proximal observations. The Statute first proposed in 2020 and approved in its final form in 2023 provides for a governance which guarantees the management of the Center operations and interaction with the Users and EO and Space Weather Communities including final users as DPC, Italian Airforce, ICAO (Fig. 9.1).

The coordinator has the task of planning and organizing the activities of the Centre and representing INGV space activities at national and international level.

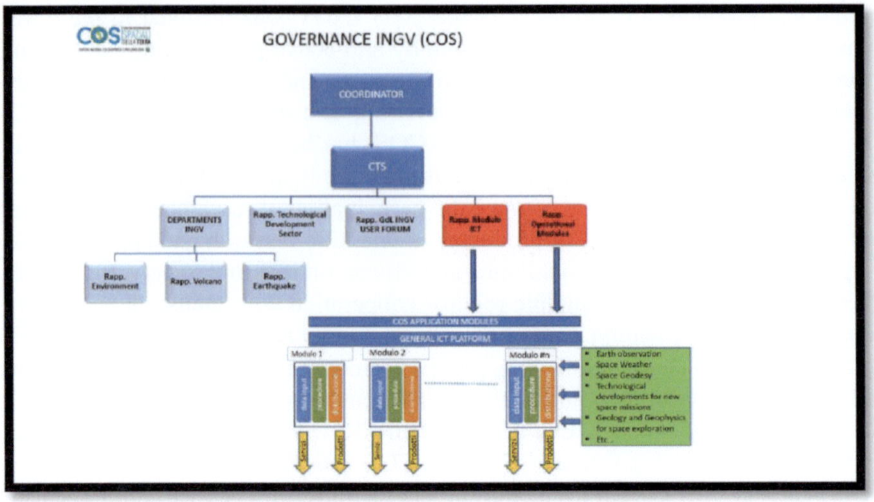

Fig. 9.1 COS organization and functions

The Technical Scientific Committee (CTS), chaired by the COS Coordinator has seven members:

A representative for each INGV Department (Earthquakes, Volcanoes, Environment) appointed by each Department Directors. A representative of the "Infrastructures and Technological Developments", appointed by the head of the INGV technological sector and three members elected internally: a representative of the INGV Working Groups at the COMINT/User Forum, a representative of the Operational Modules and a representative of the ICT-COS infrastructure. The CTS meets periodically and prepares an annual report which is transmitted to the Directors of the Departments and Sections, to the Scientific Council, to the Coordination Office for the drafting of the INGV Three-years Activity Plan (PTA, https://piao.dfp.gov.it/node/16221), to the General Director and to the President of the INGV.

The ICT-COS infrastructure represents the IT backbone of the Center and has the very important task of enabling the operation of the scientific and operational modules. Operational Modules: Each Operational Module is the fundamental unit in the COS system that allows the transfer of research activities towards products organized and usable by the scientific and applicative community, it is made up of the related personnel who propose innovations, define the requirements and carry out research and development activities aimed at the creation of future products and/or services. Developed by INGV for research activities and to support specific EO integrated services, based on INGV expertise required by external users. In the following chapters a more detailed description will be given on ICT-COS permanent module and on thematic modules which represents the first operational implementation of the COS scientific services.

9.3 COS Modules Implementation

In order to implement the COS operational structure, the first task of COS was the extensive survey on the activities developed by INGV groups working in the field of Space Observations of the Earth. To achieve this goal a web portal (https://cos.ingv.it/) was created by the INGV Graphics and Images Laboratory. At the same time, the cosnews@ingv.it group was established, which allows for rapid circulation of information from the COS-CTS to the members of the registered group, highlighted on the home page to encourage registrations to the COS group.

9.3.1 COS Operational Modules Survey

The Survey to identify the actual modules based on space data available to start the COS operational activities was performed using a specific questionnaire accessible by the COS web portal to all INGV researchers/technologists for the widest possible participation. The questionnaire was designed to find a compromise between simplicity to permit a quick compilation, and the necessity of providing details on the state of implementation and experiences of INGV colleagues in the main sectors proposed. During the survey, the compilers were given the opportunity to select a main and a secondary area of interest among the following list: Galileo, Space Weather, Space Geodesy, Earth Observations, Calibration and Validation and Payloads (satellites and airplanes). After the closing of the survey in 2021, the CTS COS decided to group the forms received in the following areas: Infrastructure Development, Space Weather, Space Geodesy, Earth Observation. Furthermore, it was defined as an additional area which included those activities related to 'technological development' (e.g. development of sensors, instruments etc.). Finally, the activities related to the calibration and validation of satellite data (CAL/VAL) were divided into two groups: those that require specific measurement campaigns (CAL/VAL on demand) and those that use instruments that provide data continuously (systematic/fixed CAL/VAL).

9.3.2 Operational Modules

The INGV internal survey gave a detailed view of the INGV space activities (Fig. 9.2) and capabilities allowing the identification of potential operational Modules and the related areas of activity in the space sector. This information was particularly useful for planning the next steps of COS activities and continuing the internal discussion.

Furthermore, through the analysis of all the information provided by the forms, a number of 'operational modules' were identified. Each Operational Module is the fundamental unit in the COS system that allows the transfer of research activities

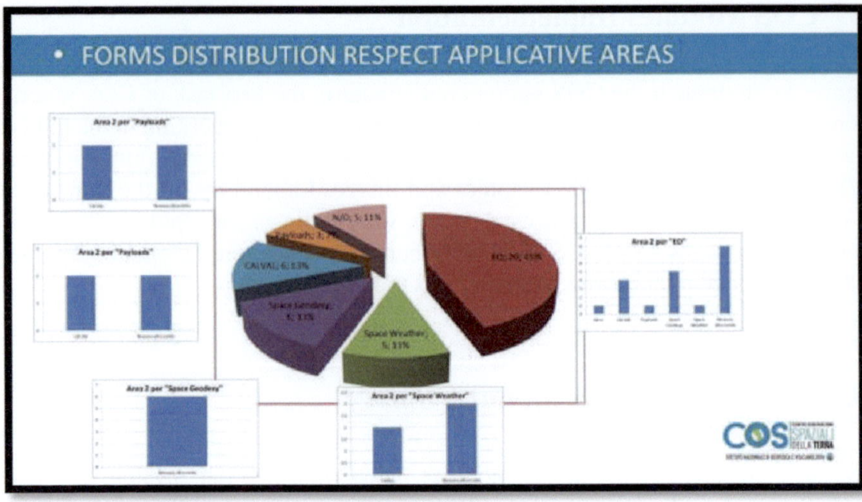

Fig. 9.2 The pie chart shows the analysis of the distribution of the areas indicated as "main" (areas 1) by the questionnaire participants. The histograms show the distribution of the Areas indicated as "secondary" (area 2) with respect to each Main Area

towards products organized and usable by the scientific and applicative community. Here are listed the COS operational modules identified in the 2021 survey and currently being integrated into the ICT-COS platform (Table 9.1).

9.4 PEOS E-Infrastructure for COS-ICT Platform

In the frame of Italian PNRR Intervention line "Research Infrastructures (IR)", defined by the MUR, INGV is leader of the MEET Project (https://meet.ingv.it/) which includes a large number of partners from Italian Research Institutes and Universities (KOM November 2022 end 2025). In this context, the COS coordinates a specific Work Package (WP 9) where an ICT platform is being developed by INGV, CNR-IREA with the technical support of specialized private companies selected by open competition in the frame of MEET Project. PEOS aims to build a common platform of data and products of Earth Observation from space, an important and challenging effort in one of the most promising scientific disciplines for hazard assessment and mitigation.

The PEOS is the ICT platform development and implementation of various COS modules described in the previous chapter is planned [5]. PEOS provides the capability to manage the import of many different space-based data into the system, it supports the generation of services suitable for society providing by means of dedicated gateway a list of generated data. This transversal ICT solution will therefore be based on the modules composing the backbone of COS itself. The processing

Table 9.1 COS Modules identified in the 2021 survey

No.	COS modules	No.	COS modules
1	Space weather for aviation—PECASUS	11	Data assimilation of volcanic ash clouds monitoring
2	Space weather service for SWESNET	12	Volcanic hazard monitoring from space
3	Technological development for aero-space systems and grounding equipment	13	MeteoSAT rapid response web service (MS2RWS)
4	GNSS archiving facility and standard geodetic products	14	Multi/hyper—spectral observations for volcanoes and environmental applications
5	Earthquake damage mapping	15	Digital surface models from EO data processing
6	SAR products for geophysical applications	16	Characterization of forest fires and burnt areas
7	Geodetic scenarios and modelling from geodetic data	17	UAS based for CAL/VAL activities
8	3D displacement field mapping from geodetic data	18	Remote ground-based observation of volcanic degassing and ash
9	Lithosphere-atmosphere–ionosphere coupling (LAIC) analysis	19	Magnetic methods for the source apportionment of PM in the urban environment
10	ASH_ICE_SO2 from volcanoes	20	Optical satellite data acquisition system—MAST infrastructure

modules are based on procedures developed by INGV and IREA. Each module has a particular objective and is composed by specific processing chains which to realize products which could support the monitoring of Earthquake, Volcano and Space Weather phenomena allowing the estimation of a number of parameters based on space data which monitor the earth surface, atmosphere in the two directions (EO satellite looking down to the Earth and ground based instruments looking toward the space as for ionosondes used for Space Weather).

The main characteristics of the COS modules are:

- The analysis of Space geodesy and SAR products are based on number of sub-modules including:
- SBAS-InSAR, InSAR-COS-INGV, Damage Mapping, 3D Displacement mapping from geodesy and Geodetic data modelling.
- The analysis of multi-spectral satellite images, the available modules allow to extract information concerning volcanic cloud properties, the 3D modelling, the surface temperature estimation and other environmental parameters, the Volcano Hazard Monitoring from Space using Artificial Intelligence and the definition of fire front during fires.

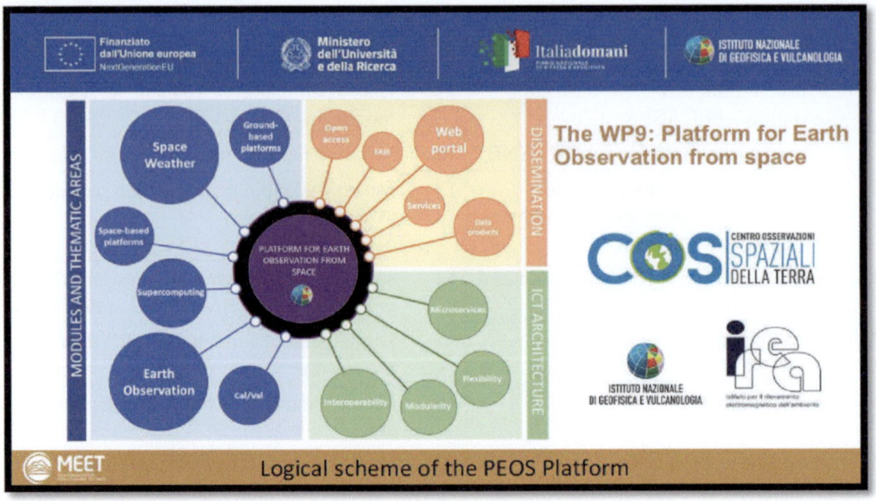

Fig. 9.3 Logical scheme of PEOS platform

- The Space Weather has 3 specific sub-modules plus the LAIC Sub-module Multiparametric data to study the atmosphere-lithosphere coupling and RF emissions
- A specific module will provide calibration and validation (Cal/Val) activities for Optical products using ground-based instruments on specific test sites developed in Italy, airborne campaign data (including UAV). These procedures will be based on independent measurements suitable to improve the accuracy of the generated products and for their validation before the distribution to operational users.

The PEOS platform not only host the operational modules but will provide an interface toward the EPOS (European Plate Observing System) ensuring the interoperability between the ICT platforms to permit an user friendly access to the products. In addition, PEOS will also offer the community and in particular to the new generations' researchers, engineers, and operational personnel of monitoring facilities a number of training modules and summer schools to support the transfer of knowledge on Space data for Geohazard Risk Reduction and Monitoring (Fig. 9.3).

9.4.1 PEOS Modules Implementation and Development

The ITC Platform developed within PEOS is based on a modular approach in order to accommodate new modules and products and interact with external infrastructures and data providers as well as the EPOS European Infrastructure (Fig. 9.4). In order to permit an high flexibility in the implementation of the scientific modules which could be available at different development stage or to permit the ingestion of external

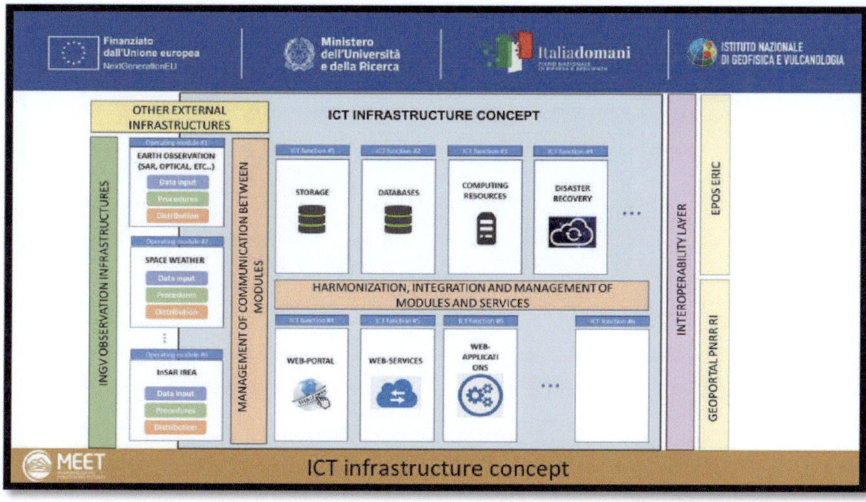

Fig. 9.4 PEOS ICT infrastructure concept

products, the ITC structure is based on External Modules (and sub modules) and Internal Modules (and sub modules).

The external modules provide products/services to the Platform by using independent processing chains and data inputs, enlarging the capability of the platform in furnishing products toward the user community. The Internal modules generate and provide products/services to the user community by using shared resources within the platform ensuring the capability of the platform to provide end-to-end processes to furnish products (Fig. 9.5).

The implementation scheme foreseen the following macro areas:

(a) Space geodesy and SAR products (Fig. 9.6); the general objective is the development of a platform for the generation, access and search of the TCS Satellite Data products of EPOS and COS to support the monitoring of earthquake induced deformation using SAR data and seismic models. The first sub-module includes architecture and development of the platform, identification of the modules to be developed, sizing of the necessary resources.

The second sub-module deals with the implementation of the modules for the on-demand analysis of satellite data (to be integrated with the platform at point 1). Therefore, development of algorithms for unsupervised processing and their integration into the platform. This module also includes the entire management part of the jobs launched by the user. The third module concerns the generation of InSAR products for the testing and validation of the platform and the population of the archive. The last module is related to interoperability with the integrated services of EPOS (ICS) and is strictly linked to the first module.

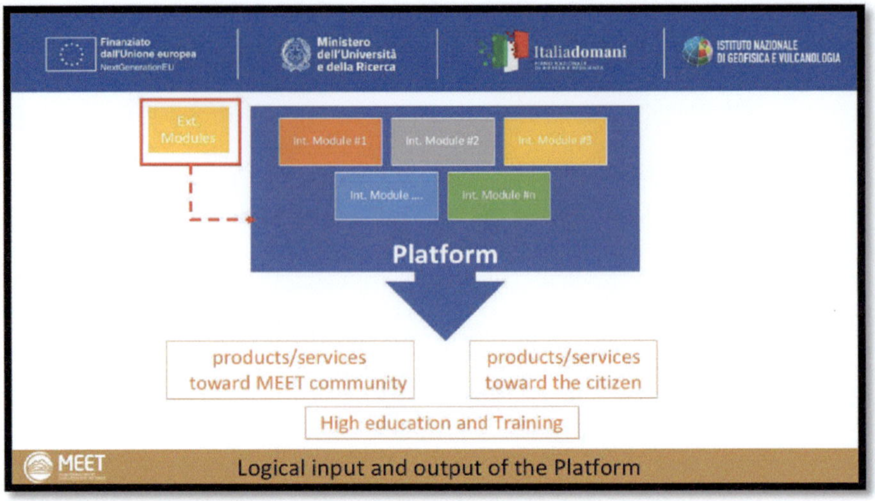

Fig. 9.5 PEOS logical scheme of data exchange

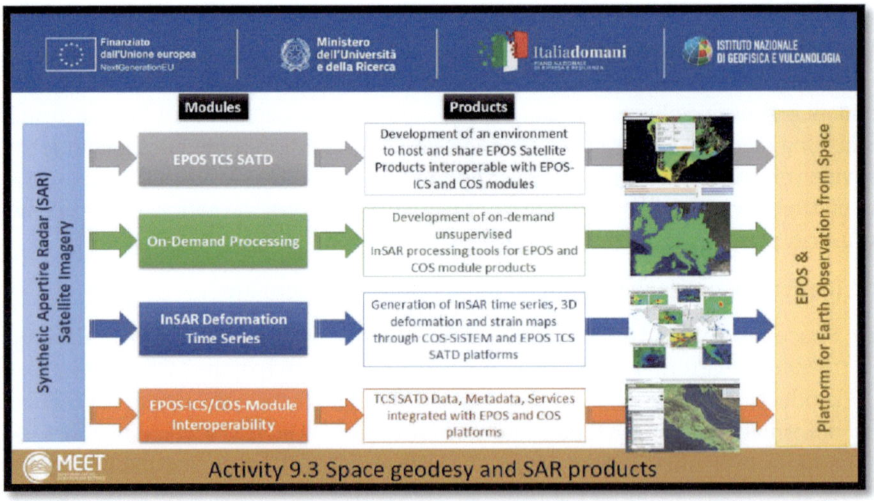

Fig. 9.6 Space geodesy and SAR products module, has the objective to develop a platform for the generation, access and search of integrated SAR products and to make them interoperable between EPOS and COS

(b) Optical data module for Volcano and Fire (Fig. 9.7); The module includes 5 sub-modules: the first is a 3D Model dedicated to the creation of the infrastructure to generate digital models of the Earth's surface (DEM) by processing high-resolution EO optical images; the second sub-module is dedicated LAND

SURFACE ANALYSIS, which includes algorithms for the estimation and monitoring of the Earth's surface temperature (Land Surface Temperature), for the calculation of environmental indices (normalized difference moisture index and normalized difference water index) and for the detection/characterization of high-temperature events such as volcanic eruptions and lava flows; the third sub-module is dedicated to VOLCANIC CLOUDS which includes products related to the detection and estimation of the height of volcanic clouds (VCH) from multispectral satellite data acquired by the SEVIRI sensor on board the geostationary satellite MSG; the fourth sub-module is for FOREST FIRE FRONT including a procedure to detect and characterize forest fires by processing hyperspectral and multispectral data; the fifth sub-module is VOLCHAZARD designed to develop a web platform for the processing of different multispectral satellite data using a flexible Machine Learning-based approach able to exploit the complementarity of optical and radar images for the detection, tracking and quantification of dangerous phenomena associated with volcanic eruptions for the timely forecasting of their impact.

(c) Space Weather and multiparametric data (Fig. 9.8): The module includes four sub-modules: the first PECASUS is already a mature service for the production of HF and GNSS space weather products to ICAO and operated in the INGV Space Weather Operational room; the second sub-module SWESNET is dedicated to the Production of Ionospheric Space Weather Products to ESA; the third module OBLIQUE SOUNDING provides information on the ionosphere even in areas not accessible to the vertical sounding, such as marine surfaces;

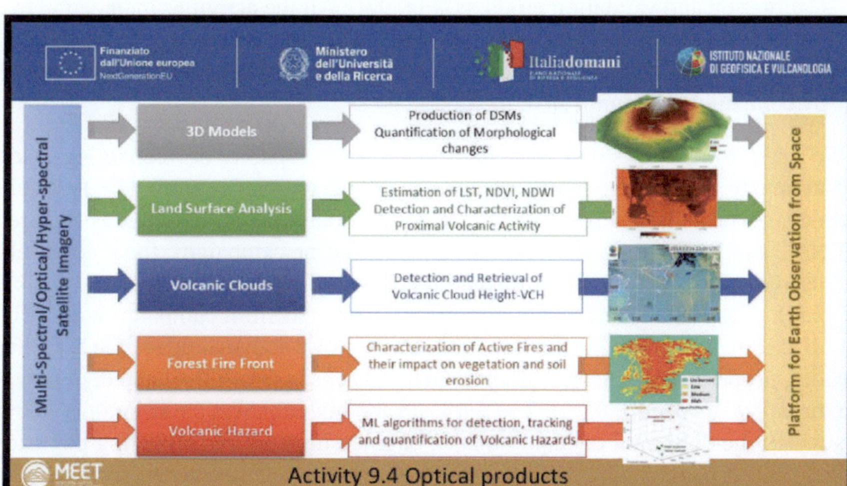

Fig. 9.7 Optical products module, include 5 sub-modules which are dedicated to volcanic monitoring, forest fires and digital terrain model retrieved by high resolution optical data

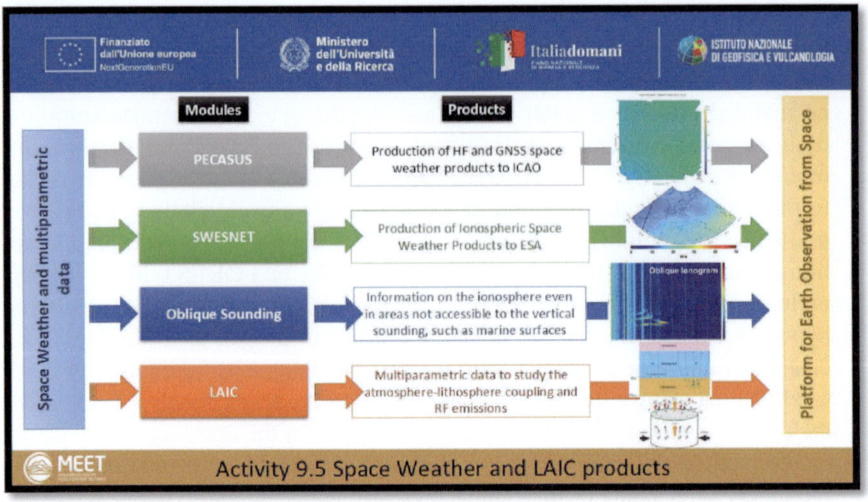

Fig. 9.8 Space Weather and multiparametric data products module has 4 sub-modules which include operational products for ICAO, ionosphere space products, and multiparametric analysis for atmosphere-lithosphere coupling

the fourth module LAIC is a Multiparametric data to study the atmosphere-lithosphere coupling and RF emissions related to earthquake, tsunami, volcanic eruption events.

(d) Calibration and validation products (Fig. 9.9): All the activities included in this module are aimed at the Calibration and Validation of satellite data (thermal, visible and hyperspectral of volcanic areas) using data collected by drone and by campaign with innovative methodologies.

We plan to use a small and harmless drone capable of recharging itself autonomously and, in the future, capable of flying without a pilot. The first sub-module support a standardized shared system to obtain the authorizations required by ENAC and EASA to fly drones and support airborne campaigns; the second sub-module includes the acquisition of data both from drone (for thermal and visible mapping), at the same time as the satellite passage, and from campaign with FTIR and LIDAR instrumentation (for the acquisition of flows and quantities of SO_2 (sulfur dioxide), properties of volcanic plumes such as backscattering flow and depolarization through UV spectrometry). The third sub-module organize the laboratory activities which the definition of requirements, engineering and implementing payloads for both aerial platforms (UAS and in future for small satellite for EO) and terrestrial system (proximal remote sensing). The technological developments will also include digital measurements, remote sensing, energy distribution on board the aerial platforms, communication and control of the on-board systems necessary for data acquisition; the fourth sub-module is dedicated to organize all data acquired from the

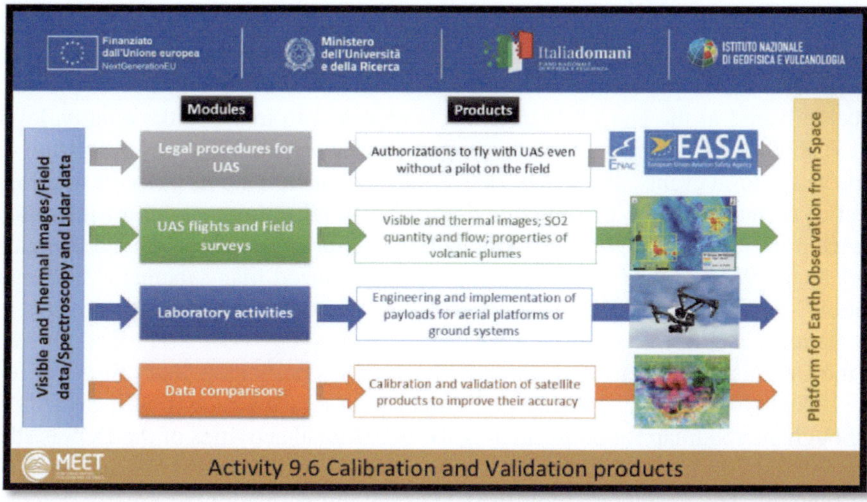

Fig. 9.9 Calibration and validation products module which include UAS systems and instruments for CAL/VAL campaign for satellite data, technological developments for ground measurements and spectral libraries

ground and from drones (as spectral libraries, temperatures distribution, gas emissions) which will be compared with those acquired from satellites to improve the precision of satellite products.

9.5 Conclusions and Future Work

The increasing in extreme weather events coupled with natural disasters events such as earthquake and volcanic eruptions increase the vulnerability of human society and infrastructures especially in developing countries. In the last 10–15 years the rapid increase of space observations provides a new and powerful tool to observe and monitor natural disasters such as extreme weather events, earthquake, tsunami, volcanic eruptions and other geohazards, both to increase the preparedness and mitigate the effects. INGV after 30 years of experience in using space data for Earth Observation and Space Weather, established a dedicated Center: The Center of Space Observation of Earth (COS). In four years of activity, COS has demonstrated a remarkable ability to coordinate INGV activities related to the Space and Aerospace sector, also managing successfully the interaction of different INGV groups working on similar sectors, thus increasing the critical mass and the expertise required at national and international level to achieve scientific objectives and to contribute to the benefit of society by supporting natural risk mitigation by means of Space data. The COS long-term objective is the consolidation of the interaction between research and industry in the sector through both cooperation activities in scientific projects funded by ASI and ESA and through direct interactions aimed at sharing scientific know-how to promote technology transfer and make research results operational. The PEOS platform developed within the PNRR MEET Project is planned to be delivered at the end of 2025 and it is expected to improve the quality, level and visibility of space-related products, particularly oriented towards mitigation of impact of natural events, and will be delivered. COS has among its tasks the development of communication, dissemination and high-level training activities internally and externally, including through participation in national events [6–8] and thematic workshops [9]. Since its establishment, it has organized many thematic "summer schools" for the training of new generations of space data experts. Participates in the National Doctorate in Space Science and Technology (DN-SST), coordinated by the University of Trento, in which 27 universities participate and offers at least one annual Doctorate scholarship dedicated to the Space sector funded by INGV. Although the roadmap for consolidating the operational activities of the COS is still long and requires large number of dedicated researchers, technologists, and systematic funding. The development of the COS ICT infrastructure has clearly demonstrated that mutual efforts between research Institutes and specialized industries of the space sector is capable to produce high quality space derived products (by means of an accurate validation of the retrieved information) that could produce specific products to support the monitoring and mitigate of Geohazards at national level and for the Mediterranean Region.

Acknowledgements We would like to thank INGV President and Board of Director to have always supported the development of COS since 2020.

We would like to thank the Italian Ministry of Research for the funding of PEOS platform in the frame of MEET project (PNRR Research Infrastructures).

We would like to acknowledge the support of SPACE IT UP Project, ASI contract n. 2024-5-E. O CUP 53D24000060005.

References

1. Copernicus Climate Change Service (C3S) and World Meteorological Organization (WMO) (2025) European state of the climate 2024. climate.copernicus.eu/ESOTC/2024. https://doi.org/10.24381/14j9-s54
2. UNDRR and WMO (2024) Global status of multi-hazard early warning systems. Geneva, Switzerland. https://library.wmo.int/idurl/4/69085
3. Centre for Research on the Epidemiology of Disasters (CRED) (2025) 2024 Disasters in numbers. CRED, Brussels, Belgium. https://files.emdat.be/reports/2024_EMDAT_report.pdf
4. Buongiorno MF, De Franceschi G, Romano V, Atzori S, Scollo S, Puglisi G (2022) INGV center for space observation of earth, URSI-AT-AP-RASC 2022, Fr-G07-AM1-3, May 29–June 3 2022, Gran Canaria
5. Romano V, Cesaroni C, Ganci G, Mainella S, Manunta M, Marcocci C, Marotta E, Musacchio M, Scotto C (2024) The PEOS e-infrastructure for upper atmosphere and Earth observation from space, URSI AT-RASC 2024, Gran Canaria, Spain, 19–24 May 2024
6. New Space Economy (NSE) European Expo-Forum (2022) 2022 Edition. Rome 1–3 Dec 2022
7. New Space Economy (NSE), European Exapo-Forum (2023) 2023 Edition. Rome 5–7 Dec 2023

8. New Space Economy (NSE), European Expo-Forum (2024) 2024 Edition. Rome 16–18 Dec 2024
9. Buongiorno MF, De Franceschi G, Romano V, Marcocci C, Atzori S, Scollo S, Corradini S, Puglisi G (2024) INGV center for space observations of earth (COS): the PEOS ICT-platform to manage integrated space products to monitor and mitigate natural hazard. In: 75th International astronautical congress (IAC), Milan, Italy, 14–18 Oct 2024

MIX
Papier aus verantwortungsvollen Quellen
Paper from responsible sources
FSC® C105338

If you have any concerns about our products,
you can contact us on
ProductSafety@springernature.com

In case Publisher is established outside the EU,
the EU authorized representative is:
Springer Nature Customer Service Center GmbH
Europaplatz 3, 69115 Heidelberg, Germany

Printed by Libri Plureos GmbH
in Hamburg, Germany